American River
Canyon Hikes

Practical Guides to Trails
in the Canyons of the
North and Middle Forks
American River

Jim Ferris
Michael Lynch
Sheila Toner

1st Edition - Dec. 2005
2nd Edition - July 2007
3rd Edition - August 2012

ISBN: 0-9772429-3-5

Introduction

This guidebook was originally designed to fill the need for comprehensive and practical descriptions for many of the best trails in the canyons of the North and Middle Fork American River. This latest edition provides four new trails (for a total of twenty-nine), along with a new centerfold area map.

To help the reader select a route for hiking, running, cycling or horseback riding, each guide provides the following: directions to the trailhead and parking, trail length and hiking time, relative difficulty, elevation change and profile, easy to read map, trail highlights, and details of what the hiker may expect to see. To facilitate reaching the start of each trail, Auburn State Recreation Area (ASRA) gate numbers and GPS trailhead coordinates are included.

Wherever possible individual trails include references to the numerical designations for the trail within the ASRA Topographical Trail Map available at Auburn State Park headquarters, the California Welcome Center, local book vendors and local outdoor stores. Note, day use parking fees apply in many areas in Auburn SRA.

We welcome your comments and any suggestions regarding this guidebook.

Frank Olrich

This book is dedicated to environmental activist Frank Olrich (1943-1999) and retired Auburn SRA State Park Superintendent Jill Dampier, who created the Canyon Keepers in 1998.

Jill Dampier

Acknowledgements

In 2003, Canyon Keeper founding member and the then volunteer chairman, Jim Ferris, initiated two projects – a series of trail guides and a self-guided trail at the American River confluence. The impetus for this book belongs to Auburn State Recreation Area (ASRA) Park Superintendent Mike Lynch, who also developed the layout and graphics design for this as well as the previous editions.

Many thanks to all the Canyon Keepers who hiked the trails, took notes and drafted descriptions for the original edition of this book. We owe special thanks and appreciation to Sheila Toner who determined all the GPS coordinates, helped edit, and personally wrote the majority of the individual guides in this edition. Rod Gross, Tom Clingenpeel and Gary Hughes were responsible for developing the elevation charts for each trail.

Jim Ferris, John Krogsrud and Laird Thompson wrote the text for Interpretive Trail Guide. We are indebted to the following for their photographic contributions: Keith Collins, Bob Griffis, Gary Hughes, Eric Peach, Margery Peterson, Gus Thomson and Sheila Toner.

Finally, we are very thankful to Mary Helen Fein, owner of **Parallax Design Group** in Auburn, who created and generously donated the trail map graphics.

Table of Contents

(* New this edition)

American Canyon Trail (#1 on the ASRA Topo Trail Map)

Distance: 2.4 miles to river; 1 hour down, 2 hours up (hiking)

Difficulty: First 1.9 mi: easy down, moderate up
Last 0.5 mi: mod. down, difficult up

Elevation Change: +/- 1,100 ft. (see below)

Trailhead / Parking: (N38-54-809; W120-55-624)

Trailhead is on Pilgrim Way east of Cool. Take Hwy 49 south to Cool. Turn left on Hwy 193 at blinking red light and drive 5.7 miles. Turn left on Pilgrim Way and look for trailhead on right side just before the gated entrance to Auburn Lake Trails. Curbside parking is available along Pilgrim Way on both sides of trailhead, and overflow parking can use private land on the left side of Pilgrim Way 100 yards before trailhead.

Description

This steep trail has everything needed for a great day of exploration: rugged mountain canyons, a beautiful stream with a waterfall, a wide variety of plant and animal life, gold rush era history, and panoramic views of the Middle Fork American River. This is an excellent hike for bird lovers and picture takers – even artists with drawing pads and paint boxes. Remember your binoculars, camera, sunscreen, water and a lunch to enjoy while you savor the scenery.

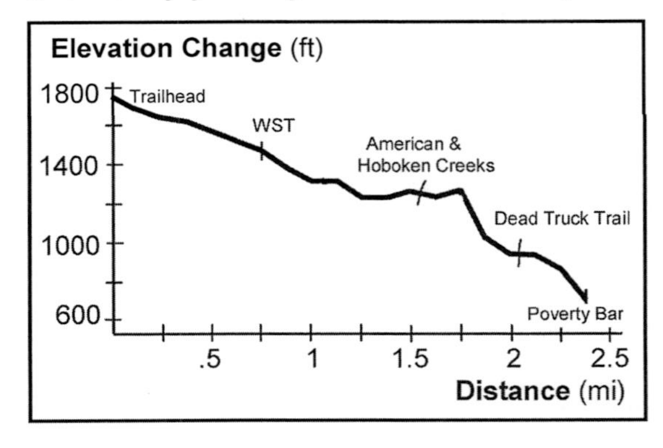

The American Canyon Trail is densely treed along the first half, mostly with various species of oak and pine. Laurel and blackberry bushes can also be found along most of the trail. Ankle-high poison oak creeps onto many parts of the trail, so please use caution. After the first mile, the trail gets more sunlight and will be hot in the summer – early morning use is recommended.

About 0.8 mile from the trailhead, the trail joins the historic Western States Trail (WST) for a short distance (see sidebar). Take a right at the first trail intersection with the WST, and go straight at the next intersection to stay on the American Canyon Trail.

At about 1.5 miles, the trail crosses American Canyon Creek, followed immediately by Hoboken Creek. These creeks join just below here. Look for California newts (a.k.a. firebellies) on the trail and in the pools.

The terrain levels out at the site of a historic gold rush town, once home to thousands of miners but now almost completely reclaimed by Mother Nature. To reach the site of this old mining town, turn right after crossing Hoboken Creek and hike off-trail for about ¼ mile upstream. The site is across the creek. Returning to the main trail, if you look closely, you may see on the right through the trees, a 30-foot high rock dam, hidden behind a pile of tailings, which was built by hand during the gold rush era.

About 1.9 miles from the trailhead, a small, very steep unmarked side trail on the left leads down to some very nice falls and a great pool where you can escape the heat on hot summer days. Please use caution while climbing on the wet, slippery rocks. This side trail to the falls is just a few feet before the intersection with Dead Truck Trail on the right.

After the intersection with Dead Truck Trail, the path becomes very steep, dropping 600 feet in less than ¾ miles on its way to Poverty Bar on the Middle Fork American River.

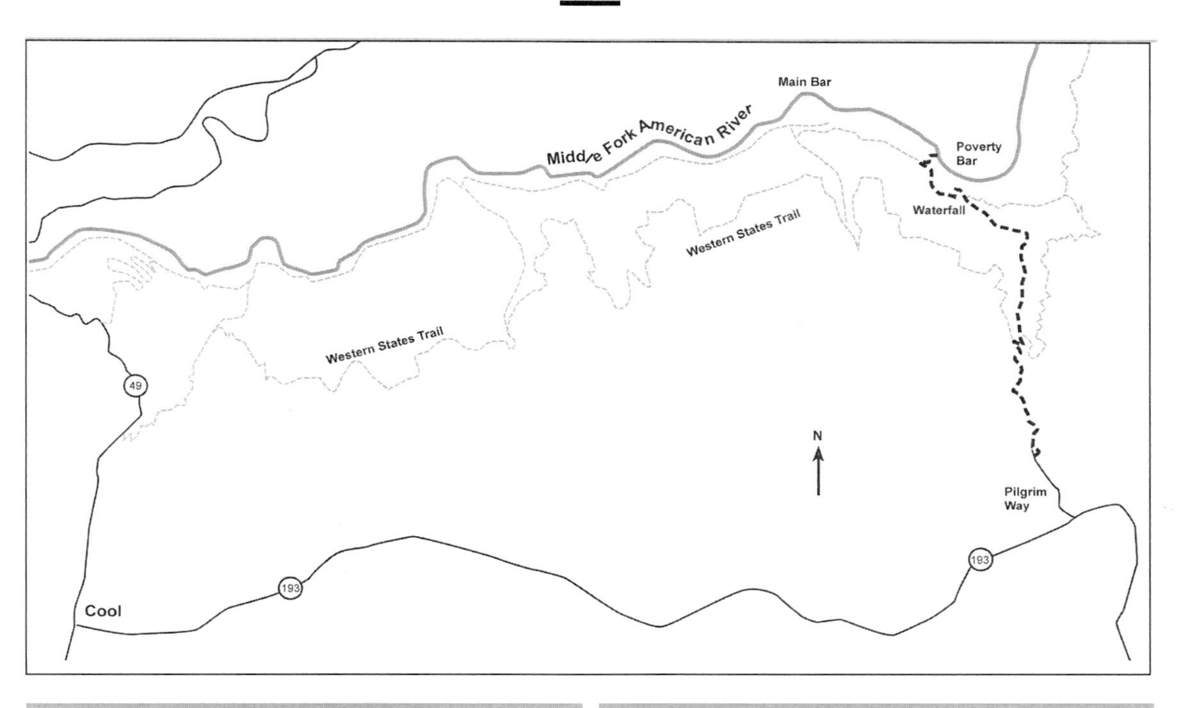

Did You Know? – The Western States Trail (WST) originally stretched from Sacramento to Utah. The Sierra Crest portion of the trail, blazed by Paiute and Washoe Indians and later used by miners, is now the route of two world-famous endurance races: the Tevis Cup Ride for equestrians, and the Western States 100-mile Endurance Run. Both races begin before dawn at Squaw Valley in the Lake Tahoe Basin and end in Auburn after nightfall, traversing roughly one hundred miles.

This part of the WST (from Ruck-A-Chucky campgrounds to the Route 49 crossing) is different for runners and equestrians. Endurance runners hug the canyon wall on a narrow, winding trail about 600 feet above the river, while Tevis Cup riders take a wider, flatter trail closer to the river, following the route prospectors used to get from bar to bar along the river. The American Canyon Trail crosses the higher WST for runners and ends at the lower WST where Tevis Cup riders cross the river at Poverty Bar.

There were extensive mining activities along the Middle Fork American River as seen in this 1858 photo of Maine Bar.

American Canyon – Dead Truck - WST Loop

Trails Included: American Canyon (#1)
Dead Truck (#10)
Western States Trail (WST)(#45)

Distance: 5 miles round trip- 2.5 hours (hiking)

Difficulty: Easy down; moderate up, except for one 0.3 mi section on Dead Truck Trail which is difficult.

Elevation Change: +/- 900 ft (see below)

Trailhead/Parking: (N38-54-809; W120-55-624)

Trailhead is on Pilgrim Way east of Cool. Take Hwy 49 south to Cool. Turn left on Hwy 193 at blinking red light and drive 5.7 miles. Turn left on Pilgrim Way. Look for trailhead on right side just before the gated entrance to Auburn Lake Trails. Curbside parking is available along Pilgrim Way on both sides of trailhead, and overflow parking can use private land on the left side of Pilgrim Way 100 yards before trailhead.

Description: *Nice loop through heavily forested canyon. At times, it feels like the Northwest not California. The scenery on the loop is more varied than found solely on the American Canyon Trail. It offers a different perspective on the canyon along with views of the Middle Fork American River. It traverses an older section of the Tahoe to Auburn trail than currently in use. The Dead Truck and WST sections offer a longer but generally flatter, more undulating and visually interesting route than the steady uphill of the American Canyon Trail.*

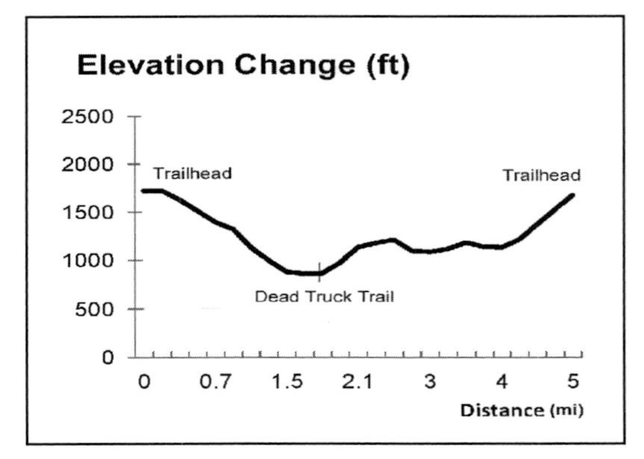

This Loop lets one see this heavily forested area, from a variety of perspectives. It offers a taste of history, along with creeks, pools, and views. It follows the American Canyon Trail (see separate guide) from the parking area, referred to as Third Gate on the trail signs, for its decent down the forested canyon. After approx 0.8 mi., take a right at the first intersection with the WST, and go straight at the next intersection to stay on the American Canyon Trail.

At about 1.5 miles, the trail crosses American Canyon Creek, followed immediately by Hoboken Creek which join just below the trail crossing on their way to the Middle Fork American River. Shortly after crossing the creeks, try to spot the 30 ft high rock dam on the right behind some mine tailings; part of the gold rush mining town of Hoboken once located above it. A small path is often visible through the foliage.

About 1.9 miles from the trailhead, there is an open area, which makes a nice break spot. The small, very steep unmarked side trail on the left leads down to a waterfall and small pool. Be careful, the rocks are slippery. A few steps beyond the side trail, the Dead Truck Trail begins on the right. (N38-56-013; W120-56-469).

Turn right and go up the initial 0.3 m steep section of Dead Truck Trail. When the trail levels out, look for the dead truck in the trees below on the left. This section offers a few views through the trees of the Middle Fork and its many gravel bars below.

Soon Dead Truck trail ends at the intersection with the WST, going left and right. Go right. In about 0.3m the trail crosses Hoboken Creek with its pleasant pools and ripples. This is a nice place to cool off. After a short uphill, the trail again levels out. Enjoy the small clearings as well as views across the canyon to the heavily forested hillside across the American Canyon gorge. Ignore the unmarked paths on the left.

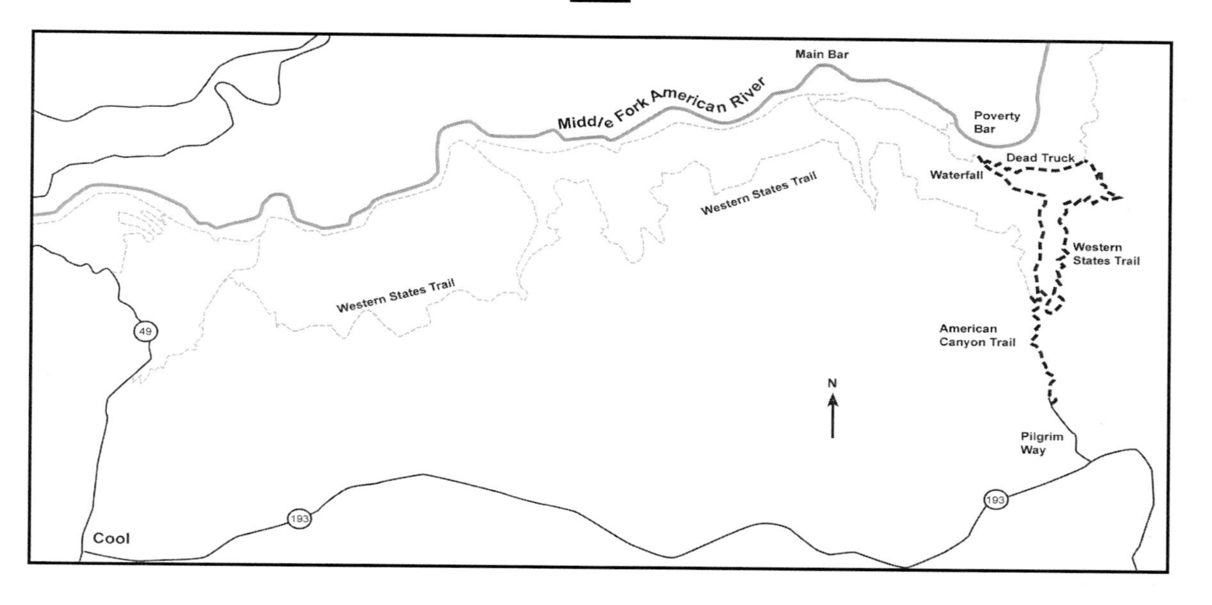

After about a mile of relatively flat terrain, the trail drops down to cross American Canyon Creek. From here a series of switchbacks leads up to where the trail joins the American Canyon Trail. At this point go left. Stay left again at the next intersection where the WST goes west. The last 0.8m to the “third gate” or parking area is a steady uphill.

Above: The dead truck that gave this trail its name.

Left: The 30 foot high rock wall just off the trail.

Applegate to Lake Clementine Trail

Distance: 2.5 miles to river one way; 1 hour down, 2 hours up (hiking)

Difficulty: Easy down, moderate up

Elevation Change: +/- 1,050 ft. (see below)

Trailhead / Parking: (N38-58-042; W120-59-056)

Parking is on Boole Rd 1.6 miles off I-80 at the Applegate Exit. At end of exit ramp turn right onto Crother Rd. Turn left at Applegate Rd (see sidebar). Go east for 0.3 miles. Just after the concrete railroad overpass turn right onto Boole Rd. Go south for 1.3 miles, passing Hilltop Ct and Roland Dr. Trailhead is an old dirt road on the left, with 4 blue posts across it, just before a big black oak with yellow road sign opposite it. There is very limited parking on Boole Rd. More parking is available on Cerro Vista, the next road on the left. A turnout is on the left at the top of the hill. The trail behind the blue posts leads in 0.3 miles back to the main trailhead.

Description

This old road provides a wide, easy route from Applegate to the waters of Lake Clementine (see sidebar). Much of the trail is in the shade of conifers and oaks. It affords a bird's eye view of the Lake Clementine beach area, as well as scenic views up and down the North Fork American River. A side trip can be made to an old lime kiln.

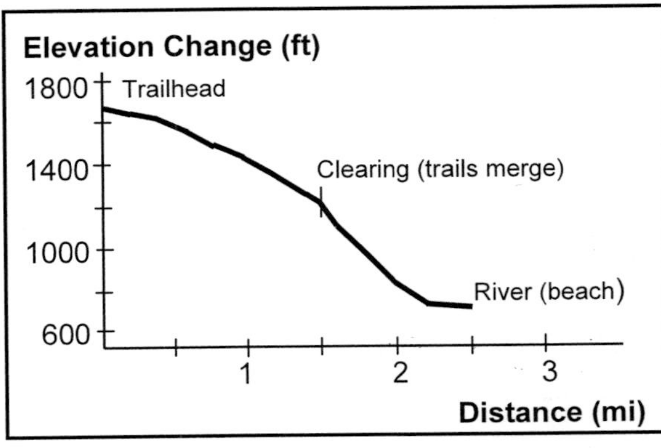

The Applegate Road to Lake Clementine Trail starts just before the black oak on Boole Rd. In less than 300 feet the dirt road reaches a small clearing. Continue straight through the clearing to the trail proper, which is a wide, nicely graded old road that drops gently down the hillside to the North Fork American River.

The trail alternates between sun and shade, with occasional glimpses of the river through the trees. At approximately 1.5 miles, a wide opening is reached where 3 trails meet. The main trail continues downhill on the left. The trail that enters at a 90-degree angle on the right meanders across the hillside to an old limekiln. A detour to the lime kiln adds 2 miles round trip.

Before continuing on the main trail, a highly recommended side trip provides some awesome views. To make this short side trip, rather than continuing down the road on the left, walk straight ahead towards the vegetation at the end of the clearing. There is a narrow path on the left, through the trees and shrubs that leads to a large limestone outcropping. Caution: watch out for poison oak. From atop this beautiful overlook, views of the sandbars in the river below and the Long Point Fuel Break Trail on the opposite ridge can be seen. The river is about 450 feet below and about a mile away. The large gravel beaches below, where the main trail will end, and the Lake Clementine beach across the North Fork, are both visible.

Back on the main trail, continue down to the beach and river. The trail is wide and generally shaded but it is steeper and not as well graded as the upper portion.

Although the trail is easy down to the river, take an opportunity to relax and cool your feet in the cold water before starting the more difficult trip back. While in mid and late summer the river is easy to ford, be very careful during the spring melt off.

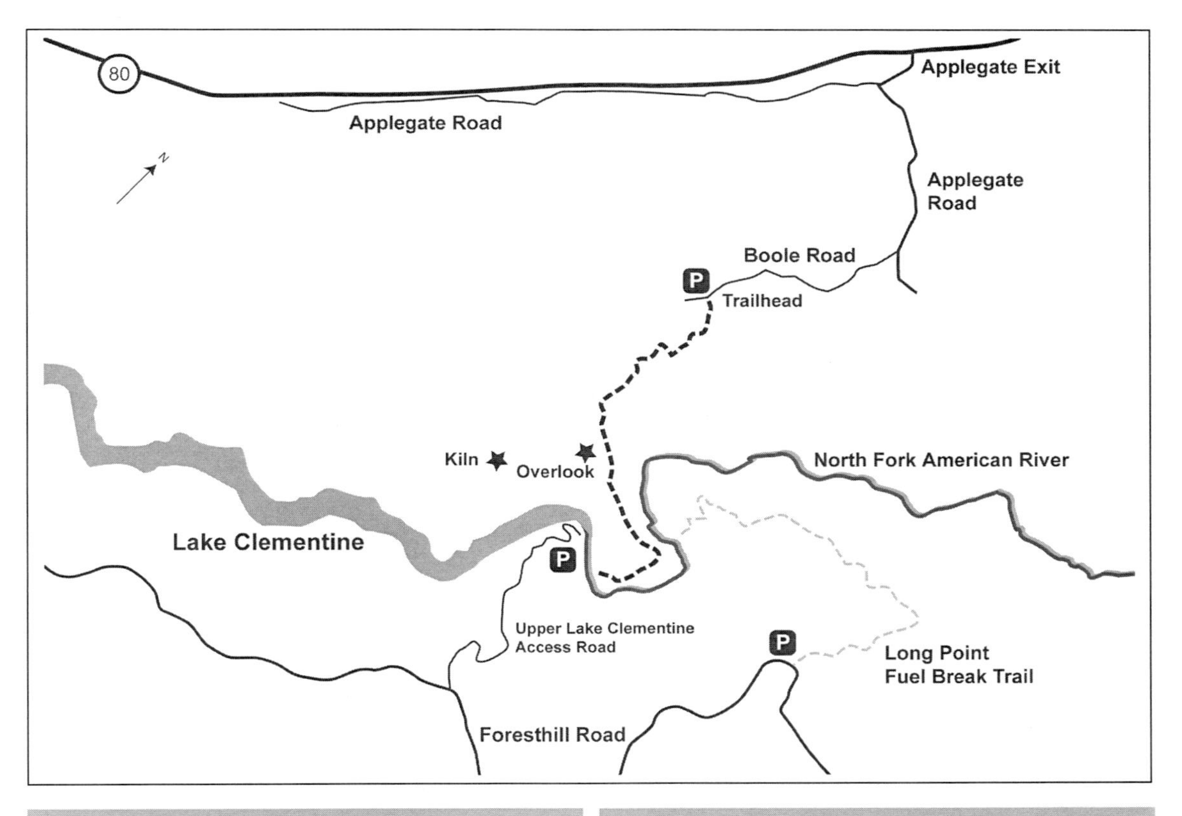

Did You Know? – Applegate Road is part of the original Lincoln and Victory Highways, which became a portion of U.S. Highway 40, now replaced by I-80. It is a scenic alternative to I-80, which can be accessed at either Clipper Gap or Heather Glen.

Did You Know? – Lake Clementine is used exclusively for public recreation. It was created by the North Fork Dam, completed in 1939. The dam was designed and built by the Army Corps of Engineers in order to collect sedimentation from upriver hydraulic mining. It was made superfluous a few years later when such operations were discontinued by state edict.

Top: A visit to the old limekiln is an optional two mile side trip.

Left: Overlook point off the Applegate to Lake Clementine Trail.

Clarks Hole Trail (#5 on the ASRA Topo Trail Map)

Distance: 0.6 miles one way; 20 min. (hiking)

Difficulty: Easy to moderate

Elevation Change: +/- 150 ft. (see below)

Trailhead / Parking: (N38-55-010; W121-02-207)

Trailhead is 1¾ miles south of ASRA Park Headquarters. Take Hwy 49 south to Old Foresthill Road at the bottom of the canyon. Continue straight for ¼ mile and park on the left. Trailhead is just beyond the parking area at the green gate (#137).

Description

This is a great family hike to a large, deep swimming hole on the North Fork American River, upriver of the Confluence. In the early 1900s, Clarks Hole (also called Clarks Pool) was managed by the city of Auburn as a municipal swimming pool, complete with concession stands and lifeguards. Today, it is still popular with Auburn locals for picnics and swim parties, although there are no lifeguards. Clarks Hole may also be accessed from the Lake Clementine Trail (see separate trail guide), which provides easier access.

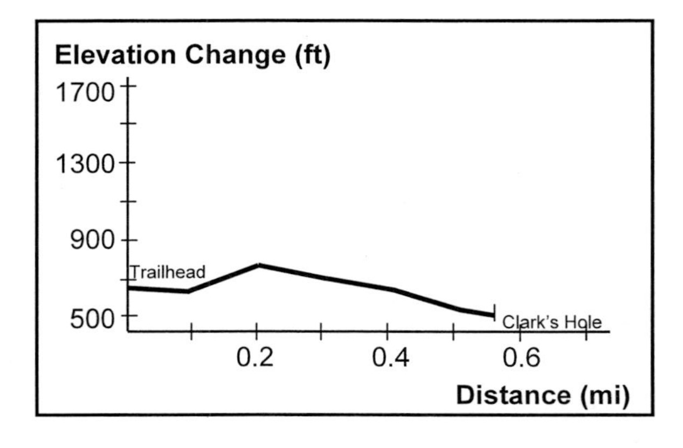

To get to Clarks Hole from the confluence, go around the green gate (#137) behind the bulletin boards and trudge up the steep fire road for 0.2 miles. Continue past the intersection with Stagecoach Trail (see separate guide) on the left, and after 40 paces, look for an unmarked, narrow trail on the right. The trail is well carved into the canyon's wall, but there are steep drop-offs to the river. Please use caution with smaller children.

The remainder of the trail gradually descends to the river under the shade of young fir trees and several species of oak. This part of the trail was once a part of the stagecoach route connecting Auburn with the gold rush camps of Foresthill, Georgetown and Greenwood in the late 19th century. The trail soon goes under the Foresthill Bridge (see sidebar).

Further along, the trail passes a small waterfall on the left, and crosses a tiny stream, both of which run during winter and spring. Just before arriving at Clarks Hole, the trail goes through a thick blackberry patch.

The trail ends with a climb down some mammoth, landmark rocks that surround the pool. The summer water temperature of the North Fork (NF) American River below Lake Clementine is surprisingly warm. It is fed by the sun-warmed water from the surface of the lake flowing over the North Fork Dam.

On the opposite side of the river, you can see the Lake Clementine Trail, (see separate trail guide) which parallels the NF American River up to the Lake Clementine Dam.

Watch out for poison oak along this trail.

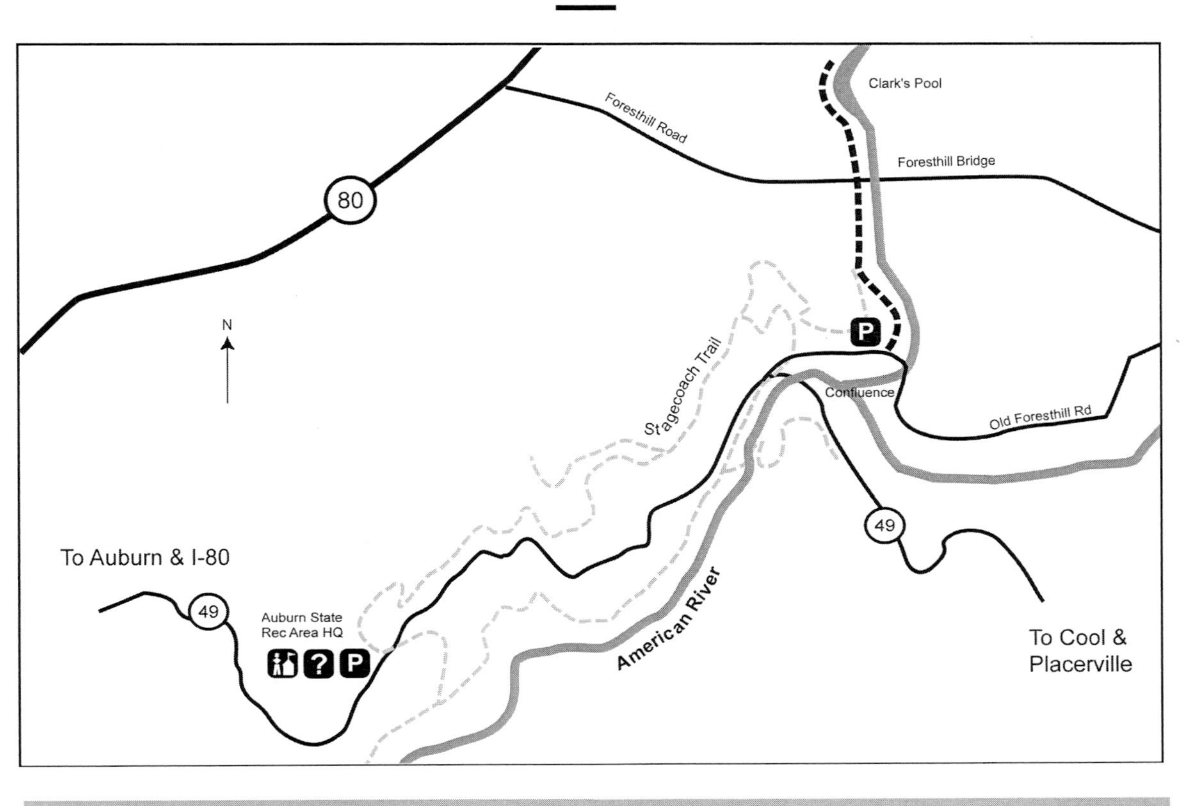

Did You Know? – The Foresthill Bridge is 2,248 feet long and connects Foresthill and Auburn. Designed to span the reservoir that would have resulted had the Auburn Dam been completed, it was opened in 1973 with much fanfare. Water was expected to reach the top of the cement piers, but today the bridge towers 730 feet above the river, making it the tallest bridge in California. It has been featured in numerous movies and commercials, and it has been the site of many stunts – both legal and illegal.

Clarks Hole.

Codfish Falls Trail (#7 on ASRA Topo Trail Map)

Distance: 1.7 miles to falls; ¾ hours

Difficulty: Easy, but trail is narrow at times with steep drop-offs to river

Elevation Change: +/- 100 ft. (see below)

Trailhead / Parking: (N38-59-995; W120-56-420)

Trailhead is on Ponderosa Way, 6 miles south of Weimar. From Auburn, take I-80 east to 2nd Weimar exit (Weimar Cross Roads) and turn right on Canyon Way. After about ½ mile, the road turns left and becomes Ponderosa Way. Past the yellow gate, the road is unpaved for the last 2½ miles. Parking is on the right just before Ponderosa Bridge over the North Fork American River. Trailhead is downstream beyond parking area. Travel in high clearance vehicles is recommended on this winding, unpaved road. Passage after winter rains may be difficult.

Description

This easy trail provides beautiful views of the North Fork (NF) American River and leads to an impressive 40' waterfall. A brochure of the flora and fauna on this trail may be available at a marker located ¼ mile from the trailhead.

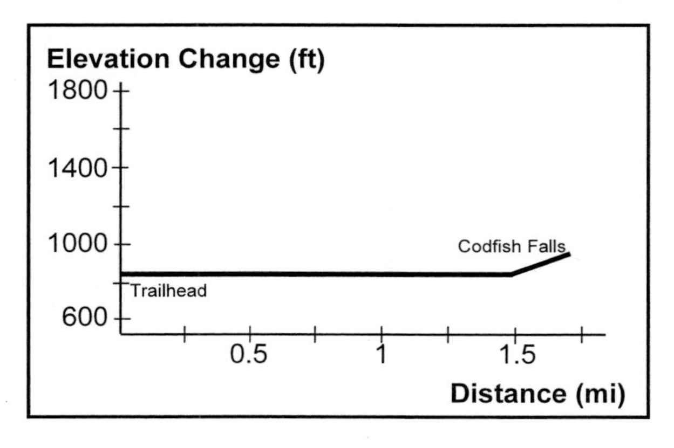

Codfish Falls Trail provides an excellent opportunity to experience some beautiful river canyon scenery. In the spring, many species o wildflowers can be spotted along the trail, and a profusion of butterflies enjoy the cool, moist morning air around the falls. In the summer, th falls can be dry and the trail can be hot in the afternoon, so morning hikes are recommended. You may want to bring a picnic lunch to enjoy a the falls. Watch for poison oak, as it thrives along the trail.

The trail starts at the parking area where Ponderosa Way crosses the NF American River. It parallels the river downstream (river-right) for the first mile and then heads up Codfish Creek to the falls. The first mile is relatively flat, and there are many side trails down to the river for those interested in cooling off on hot summer afternoons. If you look closely, you may spot dippers (also known as water ouzels) in the river. They are small, brownish-gray birds abou the size of a robin, and they can often be seen bobbing up and down, and then suddenly divin below the water. You can often see merganser ducks in the river, hunting for fish.

The North Fork American River was commercially mined for gold from the mid-19th century until well into the 20th century. Remnants of dredge mining operations can still be seen in this area. Look for the large, uniform mounds c dredged cobble tailings on the opposite bank.

Many species of plants native to a riparian woodlands ecosystem can be seen on this trail. Look for three species of oak: canyon live oak, interior live oak, and black oak. Two trees ofte confused are manzanita and pacific madrone (known for their reddish-brown bark). Both can be seen along this trail. Many fine examples of ponderosa pine, grey pine (also known as foothill, digger or whispering pine), and douglas fir can also be seen along the trail.

The area around Ponderosa Bridge is a popula swimming spot in the summer.

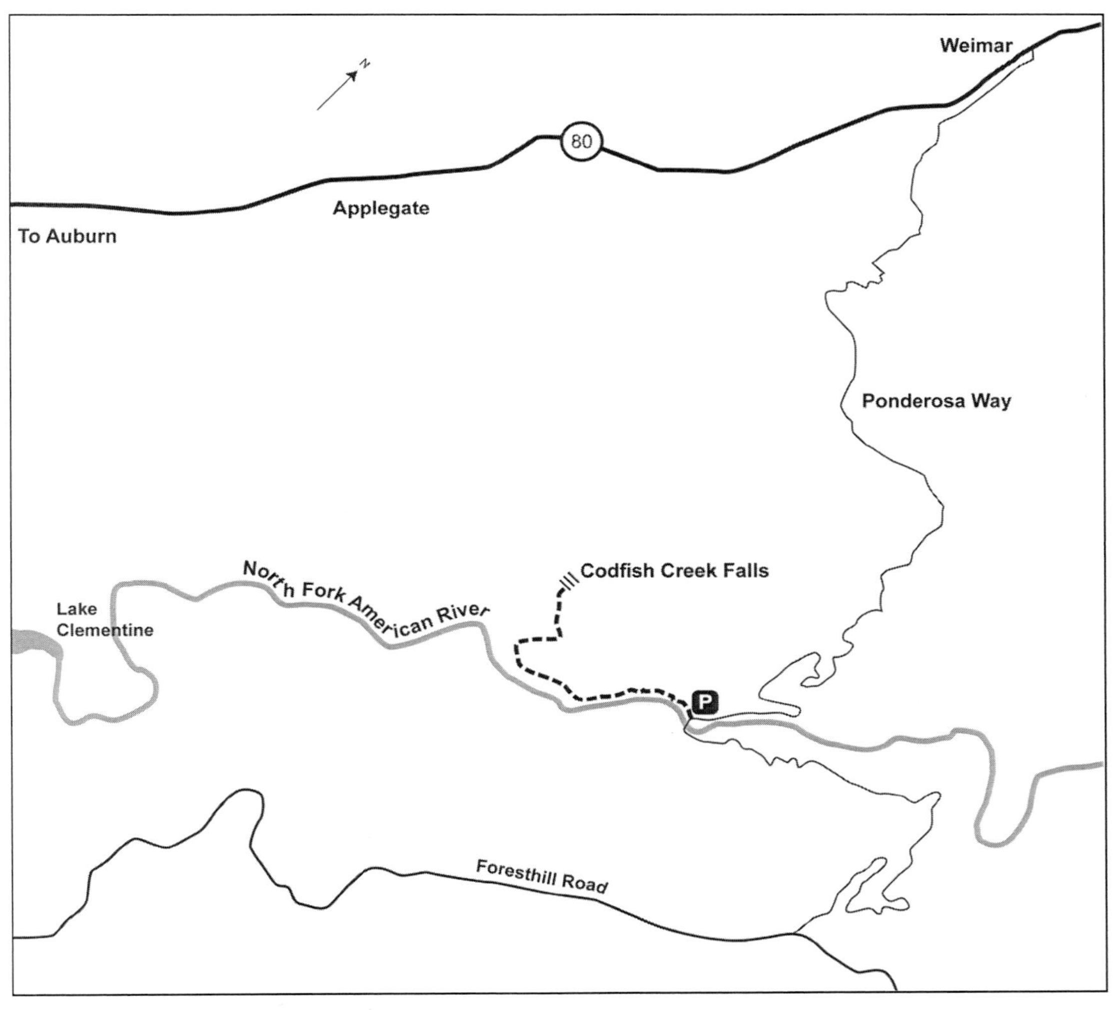

Left: Codfish Falls.

Right: An up-stream view of the North Fork and Ponderosa Bridge from the Trail.

Confluence Trail (# 8 on the Auburn SRA Topo)

Distance: 1.8 mile one way; ½ hour down; 1 hour up hiking.

Difficulty: Moderate

Elevation Change: +/- 550 ft. (see below)

Trailhead / Parking:

East: (N38-55-333; W121-00-906)

West: (N38-54-940; W121-02-150)

Trailhead (east) parking is on Old Foresthill Rd. From the Foresthill Bridge drive approx 3 miles towards Foresthill. Turn right onto Old Foresthill Road at the sign for Cool, Placerville, and Hwy 49. Drive down for approx 1.5 miles. Parking is on the left in a small dirt turnoff in front of a bulletin board, just before the entrance to the Mammoth Bar OHV area. This area may also be reached by driving approx. 1½ miles up the Old Foresthill Road from the confluence area. This turnout also serves the trailhead for the Culvert Trail that begins across the road. To reach the Confluence Trailhead, (N38-55-235; W121-00-739) walk down the Mammoth Bar Rd for 0.2 mile, past the self-pay kiosk for OHV's. Trailhead is on your right behind the yellow gate (#107).

Trailhead (west) parking is on Old Foresthill Rd at confluence area, 1¾ miles south of ASRA Park Hqtrs. Take Hwy 49 from Auburn south to Old Foresthill Road at the bottom of the canyon. Continue straight for ¼ mile. Cross the curved Old Foresthill Bridge, and park on the right. Trailhead is between large "Warning Strong Current" sign and port-a potty. At fork, go left on the trail above the beach.

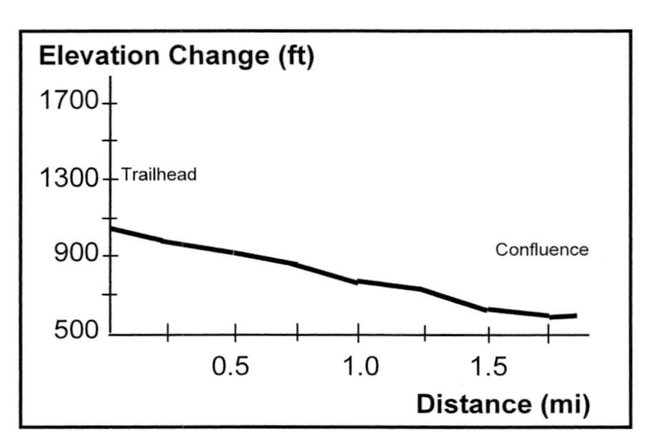

Description

This trail provides a short scenic out and back excursion through rolling chaparral and open grasslands that is dotted with wildflowers in the spring. It offers the best views of the waters of the Middle Fork (MF) American River above the confluence area, as well as unique views of the quarry site on the opposite bank. It is often combined with other trails to make a longer excursion.

The Confluence Trail offers a rolling but gentle route above the flowing waters of the Middle Fork and below the Old Foresthill Road. At one time it was paved and along the way, remnants of the old asphalt may be found. The trail is a single track with sweeping views up and down the MF. Due to its popularity with hikers and bikers going in both directions, changing road surface, dips and rises, riders often find it more challenging than hikers. Riders need to watch speed around blind curves to avoid hitting hikers. There are two small creek crossings along the trail.

Starting at trailhead east, almost immediately, there are several interesting views of the limestone quarry, located above the MF on the opposite bank, near Cool (see sidebar). Ahead there is a large limestone outcropping that is a continuation of the same limestone vein.

In spring, the trail is often awash in color. California poppies and lupine paint the hillsides with color. Wild iris, lace pods, globe lilies, Indian paintbrush, twining brodea, owl clover and other wildflowers may be spotted nestled among the trees and grasses.

There are many views of the MF as well as good views of the Quarry Trail on the opposite shore.

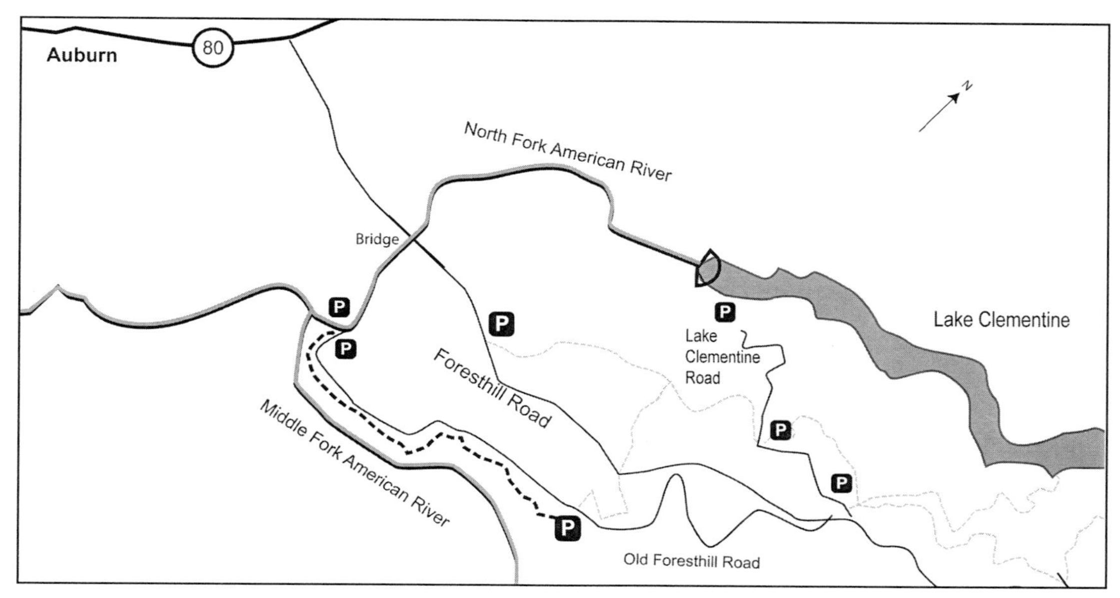

After 1½ miles, the traffic going up the hill on Hwy 49 to Cool will be seen and heard. A number of small cut offs down to the gravel bars along the river will also appear. In winter and spring the boiling white waters of the MF, as it forces its way through and over the rocks near Louisiana Bar, is an arresting sight.

Soon the trail arrives at the confluence area. In summer, a dip may be in order before you continue or head back. Several trails lead down to the gravel beach along the MF. The main trail goes straight until it heads up to the Old Foresthill Rd. and the parking area.

View of Middle Fork from the Confluence Trail.

Did You Know? – The Quarry was first known as the Cool Cave Quarry. In the 1860's Munson Manning started a limekiln here. In 1910, the Pacific Portland Cement Company began its Mountain Quarries at the site. It built a railroad and the Mt. Quarries RR Bridge to carry its products to the main line in Auburn. The Bureau of Reclamation bought the right of way and underground site of the Mountain Quarries when work began on the proposed Auburn Dam. Spreckles Sugar Company operated the quarry from the 1930's until 1987. Teichert is the current operator.

The quarry contains the largest exposed limestone deposit in the Sierra Nevada. This limestone vein runs under the MF and is seen on the opposite bank below the Confluence Trail. This same limestone vein also juts up above ground on the NF, notably as Lime Rock or Robber's Roost. The deposit is extremely pure with an average purity rating of 96-98%, which historically led to its use in the food and beverage industry, as well as in construction.

Culvert Trail (#9 on the ASRA Topo Trail Map)

Distance: 1 mile one way; ½ hour down; ¾ hour up hiking. Add 1 mile (½ hr.) from trailhead (north)

Difficulty: Moderate

Elevation Change: +/- 430 ft. (see below)

Trailhead/Parking:

North: (N38-55-400; W121-01-876)

South: (N38-55-341; W121-00-911)

Trailhead (north) parking is on Foresthill Rd, 0.5 miles east of the Foresthill Bridge, at a large parking area on the left, just after the bridge. The trailhead to the Fuel Break Trail that accesses the upper (north) end of the Culvert Trail is behind the green gate (#114) at the end of the parking strip farthest from the Foresthill Bridge. There is very limited parking at the terminus of the Fuel Break Trail in a dirt turnout off the Lake Clementine Road next to the green gate (#146).

Trailhead (south) is on Old Foresthill Rd. From the Foresthill Bridge drive approx 3 miles towards Foresthill. Turn right onto Old Foresthill Road at the sign for Cool, Placerville, and Hwy 49. Parking is on the left approx 1.5 miles from the turnoff in front of a bulletin board, just before the entrance to the Mammoth Bar OHV area. Trailhead is across the road from the parking area and bulletin board. This area may also be reached by driving approx. 1.5 miles up the Old Foresthill Road from the confluence area.

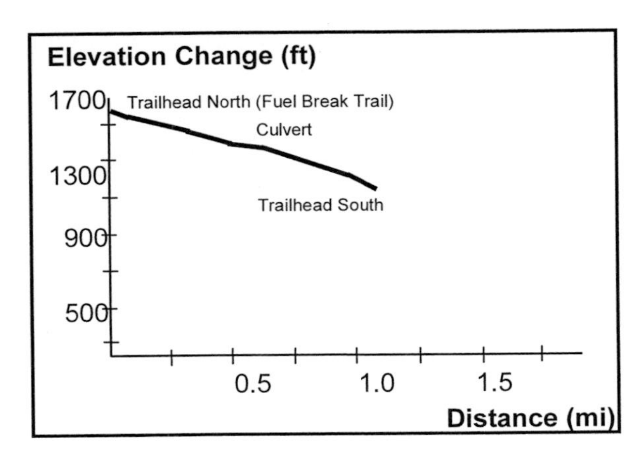

Description

This trail offers a short out and back excursion through rolling oak woodlands and open grasslands that are dotted with wildflowers in the spring. It is most frequently used to extend an outing or link with other trails. It is the safest way to traverse the busy Foresthill Rd. to go from the North Fork American River to the Middle Fork American River side of the Foresthill Divide.

The Culvert Trail is a popular year round trail. It goes from the Fuel Break Trail (see separate trail guide) to the Old Foresthill Rd and Mammoth Bar OHV area.

It is most frequently used by hikers and bikers to extend their outing by connecting with other trails, such as the Lakeview Connector, Lake Clementine, or Confluence Trail. Bikers almost universally access it from the Fuel Break Trail riding downhill. Starting from the top or trailhead (north) adds a journey of one mile on the Fuel Break Trail before reaching the Culvert Trail trailhead proper. (N38-55-834; W121-01-167). Some bikers access it from the Lake Clementine Rd terminus of the Fuel Break Trail but parking there is poor.

Starting at the north parking area, the trip begins by ascending the ridgeline via the Fuel Break Trail. After a mile, the trailhead for the Culvert Trail enters on the right. The trail is generally single file with occasional wider spots. It crosses a pretty area of open foothills woodlands dominated by blue oaks and gray pines.

Soon you will hear the sound of traffic on the Foresthill Rd. In a little less than 0.5 miles, you reach the long metal culvert under the Foresthill Rd., for which the trail is named. Expect the culvert to be muddy after rain.

After traversing the culvert, the prettiest section of the trail is reached. The path soon opens up into large rolling grasslands dotted with majestic oaks and beautiful wildflowers in the spring.

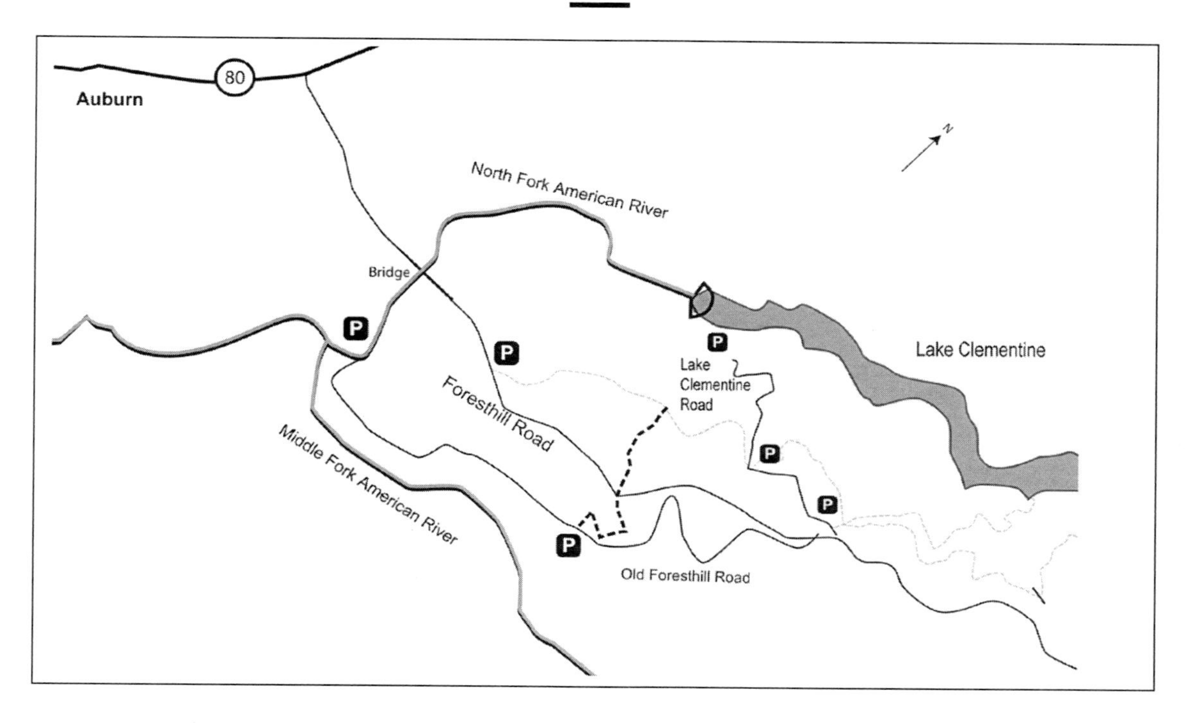

There are also vistas across the oak meadows to the forested ridgeline rising above the MF American River. As you round a corner to the right above the Old Foresthill Rd., there is a good view into the limestone quarry located above the MF American River, near Cool. It has been in operation since the 1880's.

The trail then begins its gradual descent to the Old Foresthill Rd. through a more woody area with lots of typical foothills shrubs before reaching the Old Foresthill Rd and trailhead (south).

Left: The trail is named after this culvert that runs under the Foresthill Road.

Foresthill Divide Loop Trail (#12 on the ASRA Topo Map)

Distance: 8.2 miles; 3.7 hours (hiking) Shorter loops are also possible. Add 1 mile (0.4 hrs) from Trailhead (West)

Difficulty: Easy to moderate

Elevation Change: +/- 320 ft. (see below)

Trailhead/Parking:

West: (N38-56-561; W120-58-946)

East: (N38-58-433; W120-57-253)

Trailhead (west) is on Foresthill Rd, 3.5 miles east of Foresthill Bridge, at a large parking area on the right (known locally as the Grizzly Bear House turnout). Trailhead (east) is on Foresthill Rd, 7 miles east of Foresthill Bridge (0.2 mile past Drivers Flat Road) at a large parking area on the left. Both trailheads have a porta-potty.

Description

This loop offers some beautiful views of Lake Clementine and both the North Fork (NF) and Middle Fork (MF) American River. Straddling the Foresthill Divide at elevations ranging from 1600 to 1900 feet, it provides a diversity of flora as it passes through all four American River Canyon eco-systems: yellow pine forest, foothill woodland, riparian woodland, and chaparral. It also passes through open grasslands that are abundant with wildflowers in the spring.

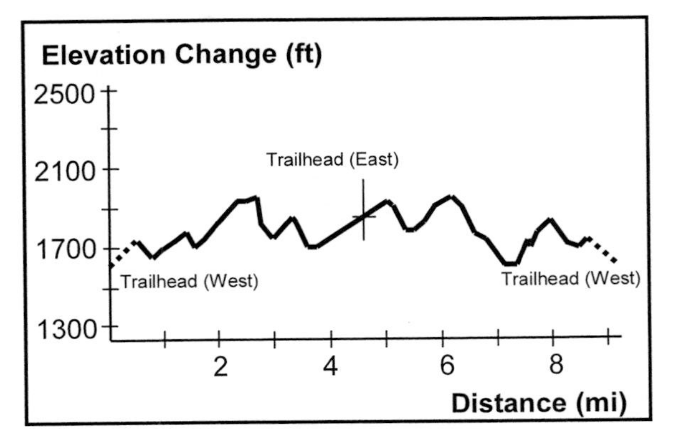

The Foresthill Divide Loop Trail offers a quintessential American River Canyon experience for hikers, bikers and equestrians, and it is extremely popular year round. Starting the loop at trailhead (west) adds one mile to the 8.2-mile loop, but it is the closer trailhead coming from Auburn.

The trail begins at a green gate (#118) behind the parking area and then turns left into the shade of oak and foothill pines. About 100 yards east of the parking area, an old rock foundation of a stage stop nestles in the trees. This hotel served miners during the gold rush era and was called the Grizzly Bear House, because the skin of an enormous grizzly bear was spread out on one of its walls, seeming to take the whole house in its embrace.

The trail soon opens up into large rolling grasslands dotted with majestic oaks and beautiful wildflowers in the spring. At 0.5 mile, turn right at the Drivers Flat marker, which puts you onto the 8.2-mile loop going counter-clockwise. At 1 mile, a side trail to the right leads to an overlook with a magnificent panoramic view of the MF American River from Poverty Bar to the east, to Brown's Bar to the west. This side trail adds 0.8 miles and 200 ft of elevation change.

Returning to the loop trail, it meanders in and out of shaded foothill woodlands ecosystem and sunny chaparral dominated by manzanita, buck brush and chemise. At 1.4 miles, the trail turns right while a dirt road continues straight.

Those interested in a shorter, 3.9-mile loop should continue straight on the dirt road; carefully cross Foresthill Road and go right for 200 yards; turn left onto Upper Lake Clementine Road and go 0.2 mi. down the road to find the loop trail again. Turn left on the loop trail and proceed 2 miles to return to trailhead (west).

For those remaining on the 8.2-mile loop, the trail becomes narrower and more scenic. At the 2.2-mile point, there is a grand view of the Middle Fork and the Cool limestone quarry to

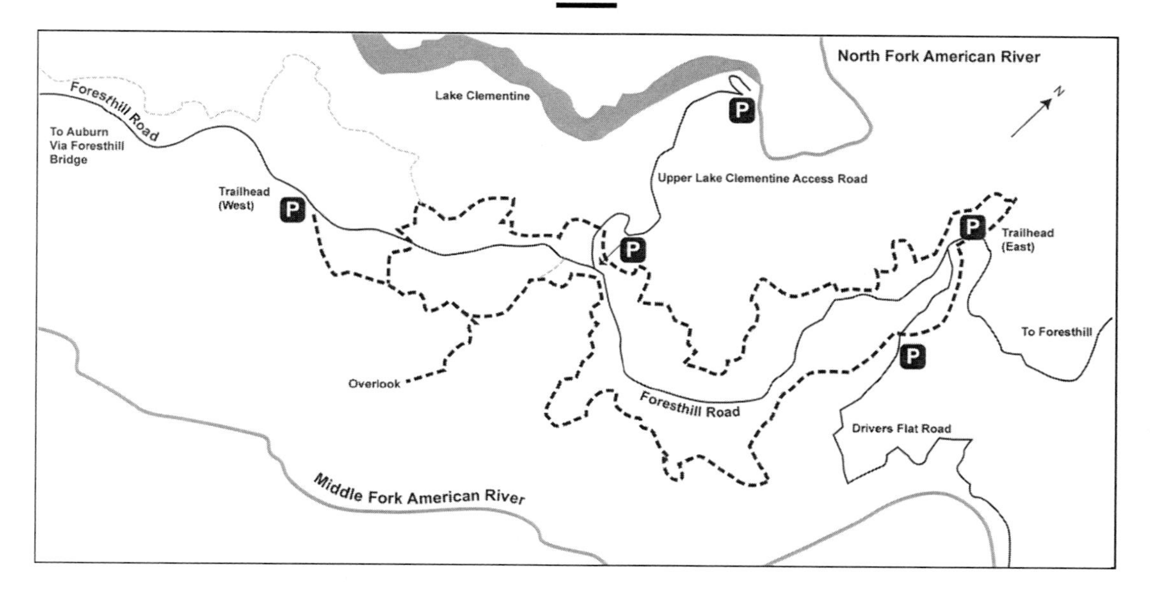

the west. At 3 miles, a well-built wooden bridge crosses a creek (running in winter and spring). A bit further, there is a fork in the trail. Stay to the left. At 4.1 miles, either continue straight on the loop trail or turn left and proceed up Drivers Flat Rd. to Foresthill Rd. Either way, when you get to the road, carefully cross (preferably when no cars are coming) and proceed to the parking lot at the trailhead (east). This marks the halfway point of the 8.2-mile loop trail.

From the trailhead (east), the loop trail continues behind the parking area. It is mostly shaded with no open meadows, passing through dense yellow pine forest and foothill woodlands. The views are less grand on this side of the Foresthill Divide, although glimpses of the NF American River and Lake Clementine can sometimes be seen through the trees in the canyon below.

At 1.5 miles from trailhead (east), there is a great view of Lime Rock (see sidebar) on the far side of Clementine Lake. At 2.5 miles, the loop crosses Upper Lake Clementine Road.

Those starting at trailhead (east) and looking for a shorter, 6-mile loop: turn left and climb Upper Lake Clementine Rd to Foresthill Rd.; turn right and go 200 yards; carefully cross the road and continue on a dirt road behind the green gate (#119); go 150 yards and turn left on the loop trail; proceed 3.3 miles to trailhead (east).

For those remaining on the 8.2-mile loop, cross Upper Lake Clementine Rd, go 3.8 miles through densely shaded foothill woodlands, and emerge on Foresthill Rd. Carefully cross the road and go 0.2 miles to a trail intersection. To return to trailhead (west), turn right and go 0.5 miles. To return to trailhead (east), go straight for 4.2 miles.

Did You Know? -- Lime Rock is locally known as Robber's Roost. When Foresthill was still a booming gold mining site, a local band of outlaws, the Gassaway Gang, is reported to have used the rock as a strategic lookout point to spot and then signal when the stagecoach was enroute from Foresthill to Auburn.

A view of Lime Rock or Robbers Roost above Lake Clementine.

Fuel Break Trail (#14 on the ASRA Topo Trail Map)

Distance: 1.5 miles One way; ¾ hour (hiking)

Difficulty: Easy

Elevation Change: +/- 450 ft. (see below)

Trailhead/Parking (N 38-55-400; W121-01-876)

Trailhead is off Foresthill Road just past the Foresthill Bridge.
Take I-80 to the Foresthill Exit. Drive towards Foresthill, and in less than a mile, cross the Foresthill Bridge. Just after the bridge, the trailhead parking area is visible on the left side of the road. The trailhead is behind the green gate (#114) at the end of the parking strip farthest from the Foresthill Bridge.

There is very limited parking at the terminus in a dirt turnout off the Lake Clementine Road, next to a green gate (#146).

Description

This wide easy trail passes oak meadows, foothill chaparral and woodlands and provides several nice ridge top views. It connects with the Lake Clementine Rd, the Culvert Trail and the Lakeview Connector Trail. It is often used in conjunction with them for longer excursions.

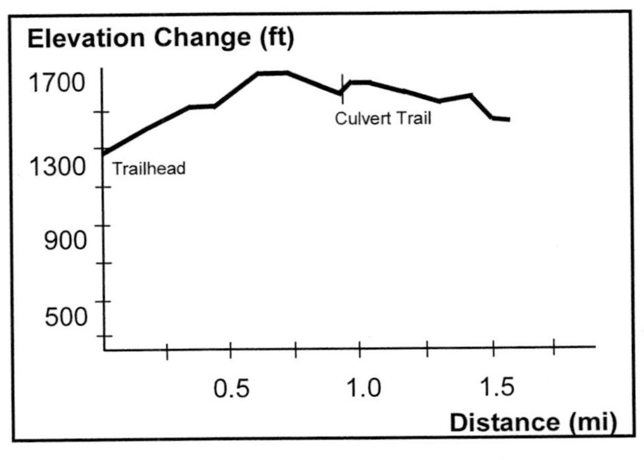

The Fuel Break Trail begins in the woodland meadow behind the green gate (#114) at the end of the parking strip farthest from the Foresthill Bridge (see sidebar). Before or after the hike, walk out onto the bridge for a bird's eye view up and down the North Fork American River.

The trail begins by climbing to the ridgeline. The steepest portion of the hike is the first ½ mile. After reaching the ridgeline, the trail levels off and then begins a gradual descent, to the paved Lake Clementine Road.

As you reach the ridgeline, look right across the Divide to the ridgeline opposite, which is above the Middle Fork American River. The limestone quarry, near Cool, which has been in use since the 1880's, is clearly visible.

Several small oak meadows are passed along the way. These areas host nice wildflower displays in the spring.

In addition to hikers and bikers, the trail is popular with the local wildlife. They frequently use it in the early mornings and evenings, especially when the manzanita trees that line several areas of the trail are loaded with fruit. They are seldom seen but their tracks show that this is a busy roadway for both large and small animals. This is an ideal place to practice reading the signs of their passage.

After a mile, the Culvert Trail (see separate trail guide) branches off to the right. A few other paths will branch off along the way for short distances. Stay on the main trail that continues straight ahead.

The trail ends at the paved Lake Clementine Road, on the left, by the large boulders marking the end of the trail. From here, two short cutoffs head toward the Foresthill Road.

The Lakeview Connector Trail (see separate trail guide.) begins a few feet across and up the paved Lake Clementine Road.

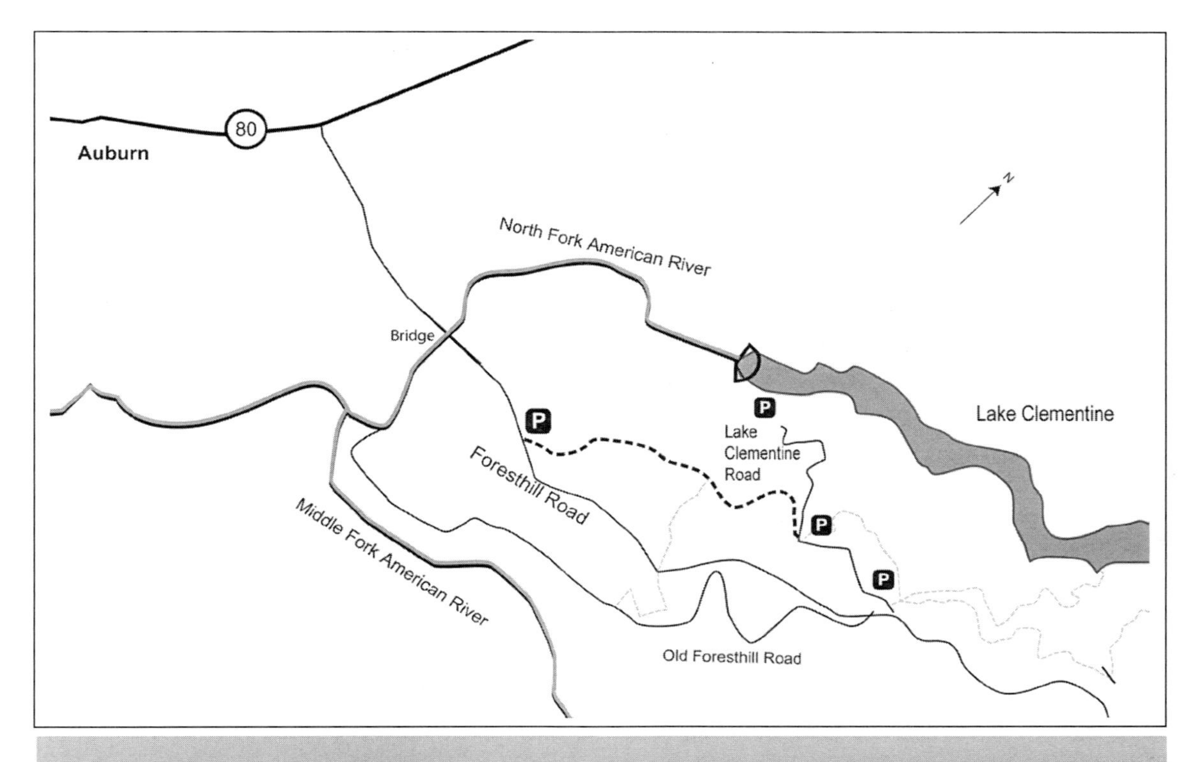

Did You Know? – The Foresthill Bridge is 2,248 feet long and connects Foresthill and Auburn. Designed to span the reservoir that would have resulted had the Auburn Dam been completed, it was opened in 1973 with much fanfare. Water was expected to reach the top of the cement piers, but today the bridge towers 730 feet above the river, making it the tallest bridge in California. It has been featured in numerous movies and commercials, and it has been the site of many stunts – both legal and illegal.

Above: Be sure to take in the view from California's highest bridge near the Fuel Break Trailhead.

Left: Bikers enjoying the Fuel Break Trail

Indian Creek Trail (#15 on the ASRA Topo Trail Map)

Distance: 2 miles one-way; 1 hour hiking

Difficulty: Easy, but the trail is narrow with steep drop-offs to the river.

Hardest portion is reaching the trailhead on the west side of Shirttail Creek.

Elevation Change: +/- 50 ft. (see below)

Trailhead / Parking (N39-02-410;W120-54-189)

Trailhead is off Yankee Jims Road on the North Fork American River near Shirttail Creek.

Take I-80 to Colfax. Exit at Colfax and turn right (west) onto Canyon Way. After approx. 0.8 miles, turn left (south) onto Yankee Jims Road. The road will quickly become gravel. Yankee Jim's Rd. is narrow, twisting, and without guardrails. It descends rapidly to the North Fork of the American River. Drive slowly. Hikers, rafters and miners frequently use the road, and visibility around the narrow curves is limited. For those not driving, there are some great views down the cliff on the right. At the river, there is limited parking on either side of the 1930 single vehicle suspension bridge. Cross the bridge, by car or on foot depending on parking availability.

After crossing the bridge, look for rock stairs down to the North Fork American River on the left. The first set of stairs is wide but uneven. Another set of stairs, not as wide but generally in better shape, is just ahead around the bend.

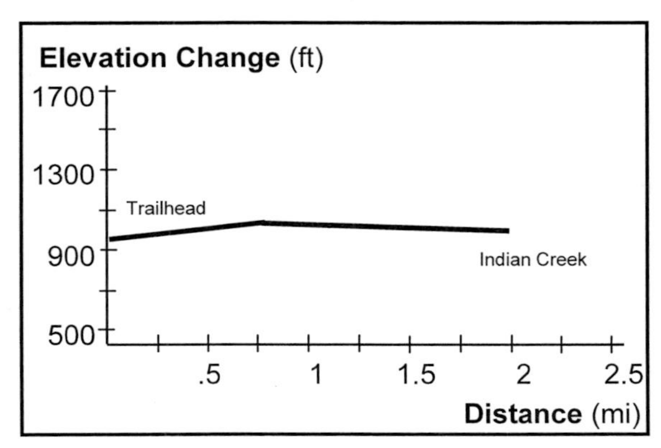

The trail proper is on the other side of Shirttail Creek. Depending on water level, wade or boulder hop across Shirttail Creek. Once on the far side of the creek you will see the trail up the bank from the river.

DESCRIPTION

This easy trail offers a fun summer hike. It allows you to enjoy the American River Canyon without having to hike in or out. It offers great views up and down the canyon as well as many opportunities to take a dip. Canyon live oaks and other foothill trees keep much of the trail in the shade, which is welcome in summer. You are seldom out of sight or sound of the river.

Indian Creek Trail goes from Shirttail Creek to Indian Creek. This narrow but easy trail is cut into the hillside above the south bank of the North Fork American River. While there are some ups and downs, there is little elevation gain or loss. It offers many nice views of the river canyon. The hardest portion is at the beginning where Shirttail Creek must be crossed to reach the trail proper.

The trail is unmarked but once across Shirttail Creek, it is visible just up the bank from the river.

Several side trails drop down to the river along the way. Whenever there is a choice, on your way to Indian Creek, always choose the right or uphill fork since the downhill choice always leads to the river. The slopes are steep so please stay on the trails.

The trail crosses two tributary streams, which are easily crossed. The first is usually dry in summer. Shortly after this, the trail crosses the larger Salvation Ravine with its almost year round water flow, just prior to arriving at Indian Creek and the official end of the trail.

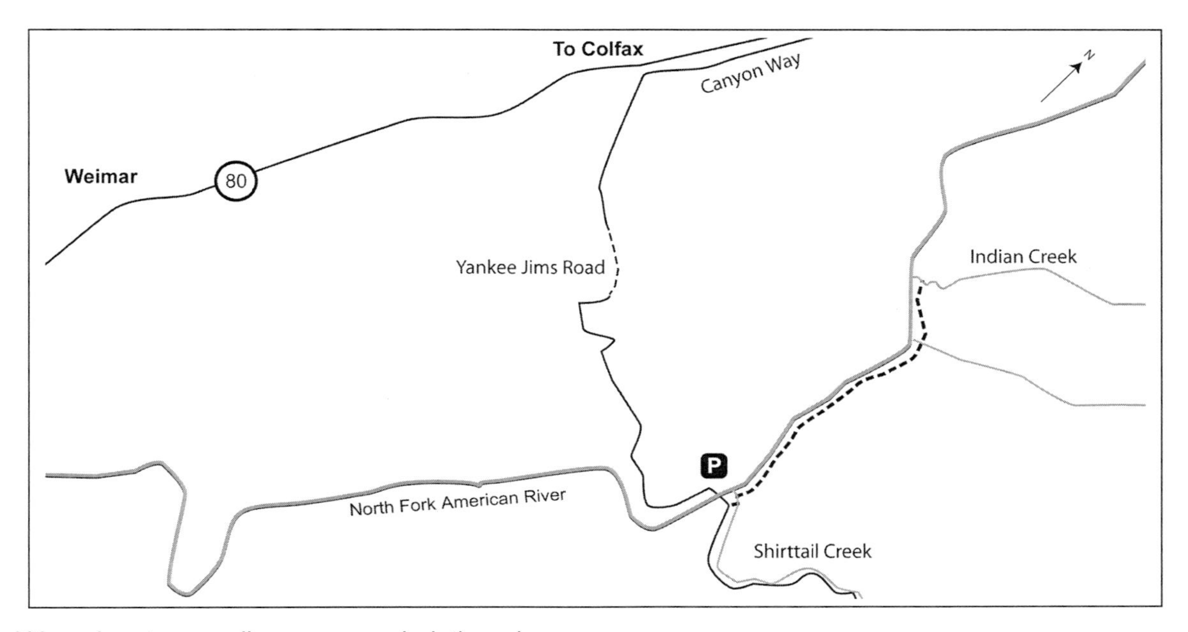

Although not generally recommended, there is a small, steep, overgrown and challenging track that leads upstream to a large pool fed by a double waterfall. This is a nice photo spot.

Time of the year is important on this trail. If you go too early in the spring and too much water is flowing, access is a problem. If you go too late in the fall, the waterfalls are a disappointing trickle.

This trail is one of the few that provide a fun river canyon experience without having to hike up or down the canyon.

Above: The Colfax-Foresthill Bridge on Yankee Jims Road.

Right: Bridge from near the start of the Indian Creek Trail from Shirtail Creek.

Lake Clementine Access Trail (#17 on the Auburn SRA Topo Trail Map)

Distance: 1.5 miles one way; ¾ hrs. (hiking)

Difficulty: Easy down, moderate up

Elevation Change: +/- 750ft. (see below)

Trailhead/Parking (N38-56-107; W121-00-254)

Trailhead is on Lake Clementine Road off Foresthill Road. Take I-80 to the Foresthill exit. From the light proceed towards Foresthill for approx 3.2 miles. Turn left on the Lake Clementine Road. Shortly after turning, the parking area is on the right. The trailhead is behind the green gate (#115) about 200 feet back up the road.

Description

This is a good year round trail due to its wide, gentle gradient and alternating stretches of sun and shade. This wide pleasant trail traverses typical foothills woodlands while providing superb views of Lake Clementine and Lime Rock as it wends its way down to lake level. It is the only trail that goes down to the lake above the dam and below the Upper Lake Clementine day use area.

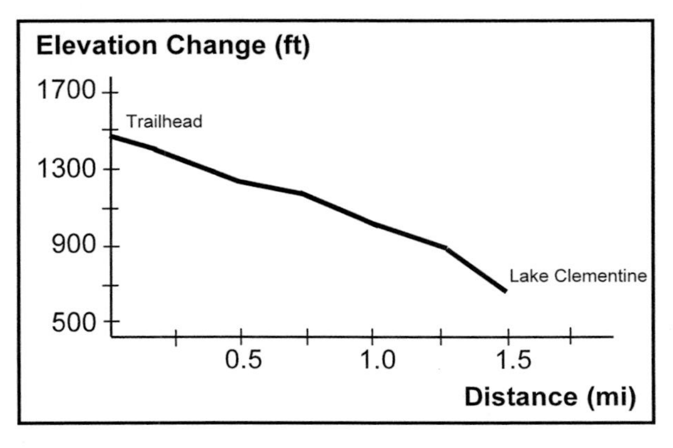

The Lake Clementine Access Trail begins behind the green gate (#115) just 200 feet back towards the Foresthill Road. The trail goes straight down the hill behind the green gate. **Do not** take the unsigned Lakeview Connector Trail (See separate trail guide) that crosses the Lake Clementine Access Trail going left and right parallel to the Lake Clementine Rd.

This nicely graded service road, which is sometimes referred to as Doc Gordon's Road, offers an enjoyable hike down to the edge of Lake Clementine. It alternates between sun and shade while offering a few aerial type views of the lake. Note the many different types of trees found along the way. There are several stands of manzanita and madrone, whose reddish trunks always feel cool to the touch, as well as ponderosa pine, bay, several types of oaks and other foothill favorites. There is a nice display of maidenhair and bracken ferns along the side of the trail about half way down in the wetter months.

The trail ends in a small clearing at the edge of Lake Clementine. A small stream, on the edge of the clearing, bubbles over moss covered rocks in this shady riparian zone. Trees overhang the area and provide welcome shade on sunny days. In summer, this is a nice place to cool your feet, while enjoying the shade and lake views. The *SS Relief* (the port-a-potty for boaters) is usually moored in the center of the small cove.

The trail offers some great views of Lake Clementine and the limestone outcroppings on the opposite shore. These outcroppings are a continuation of the same type of hard limestone deposit that has been mined for commercial use since the 1880's at the quarry located on the Middle Fork American River. The most predominant feature is Lime Rock, known locally as Robber's Roost. (See sidebar)

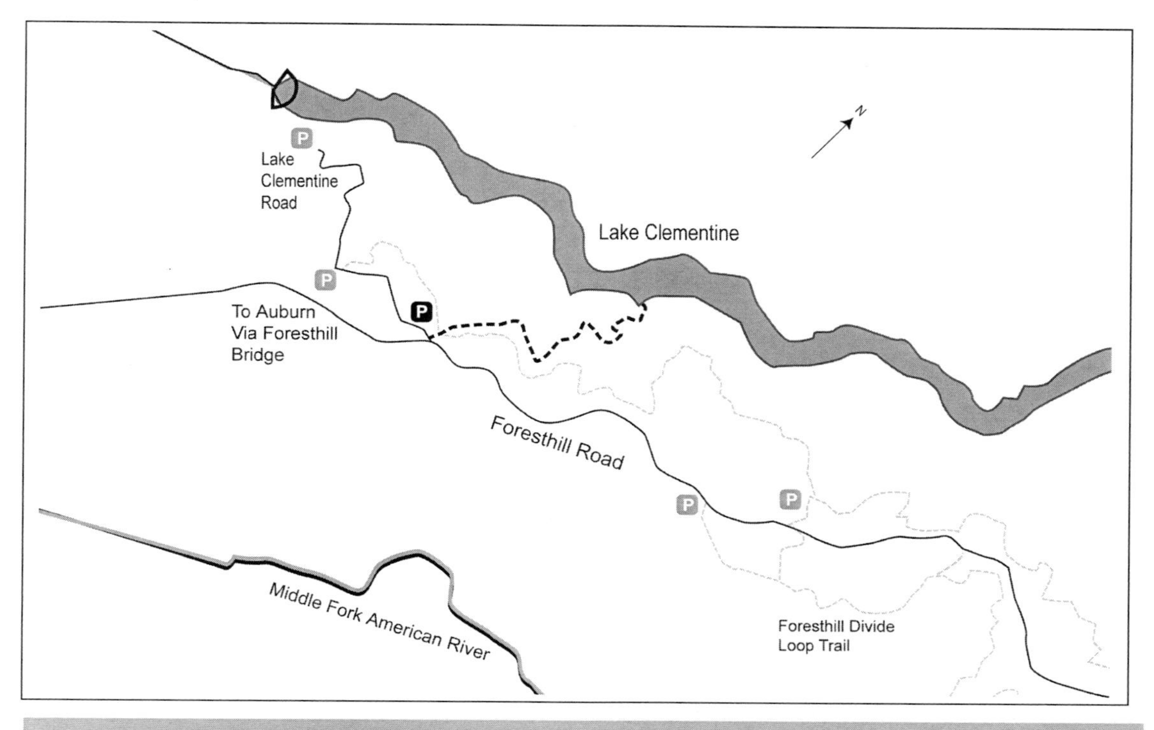

Did You Know? -- Lake Clementine is used exclusively for public recreation. It was created when the North Fork Dam was completed in 1939. The dam was designed and built by the Army Corps of Engineers to collect sedimentation from upriver hydraulic mining. It was made superfluous when such operations were discontinued by state edict.

Did You Know? -- Lime Rock is locally known as Robber's Roost. When Foresthill was still a booming gold mining site, a local band of outlaws, the Gassaway Gang, is reported to have used the rock as a strategic lookout point to spot and then signal when the stagecoach was enroute from Foresthill to Auburn.

Above: Lake Clementine offers many water recreation opportunities.

Left: Lime Rock reflected in the waters of Lake Clementine from the Access Trail.

Lake Clementine Trail (#18 on the ASRA Topo Trail Map)

Distance: 1.9 miles; 1 hr each way (hiking)

Difficulty: Easy, except for short stretch where trail narrows due to wash out

Elevation Change: +/- 350 ft. (see below)

Trailhead/Parking: (N38-54-941; W121-02-144)

Trailhead is at confluence area, 1¾ miles south of ASRA Park Headquarters. Take Hwy 49 from Auburn south to Old Foresthill Road at the bottom of the canyon. Continue straight for ¼ mile, cross the curved Old Foresthill Bridge, and park on the right. Trailhead is on the left across from the parking area behind the green gate (#139).

Description

This easy trail is ideal for a panoramic hike along the North Fork (NF) American River, much of it in the shade of conifers and oaks and in close proximity to riparian flora and chaparral. Several side trails lead down to the river, including one to the ever popular Clarks Hole, and the last one descending to a deep pool beneath the North Fork Dam where spectacular views of water cascading over the dam can be seen.

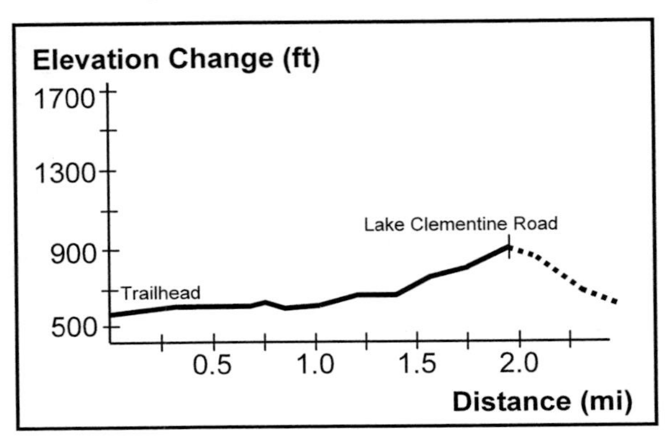

Lake Clementine Trail begins at the confluence area on the far side of the curved Old Foresthill Bridge (built in 1955). It parallels the NF American River upstream, river left. At ¼ mile, concrete abutments for what was known as the Steel Bridge (1911-1955) may be seen on the opposite riverbank. At ½ mile, the trail goes under the Foresthill Bridge (see sidebar).

At ¾ mile, Clarks Hole can be seen on the left. It is a deep and clear rock-lined swimming hole that has been popular with locals for over one hundred years. A short side trail, on the left, leads down to the pool and a sandy beach. The summer water temperature of the river below Lake Clementine is surprisingly warm since it is fed from sun-warmed water from the surface of the lake flowing over the North Fork Dam. By contrast, the Middle Fork American River is fed from the bottom of Oxbow Reservoir and is considerably colder.

At 1 mile, the rock abutments from a wooden covered toll bridge, built in 1875 and used until 1911, are visible on the opposite riverbank. A short distance upriver, if you look carefully, you can see evidence on the opposite riverbank of abutments where three different bridges existed between 1852 and 1875.

Continuing upriver, this part of the trail uses a largely shaded roadway that was once the old stagecoach route connecting Auburn with the gold rush camps of Iowa Hill, Yankee Jims and Foresthill in the late 19th century (see sidebar).

The Lake Clementine trail ends at Lake Clementine Road. Follow this paved road to the left for about ¼ mile and then take the unmarked side trail on the left towards the river for an exciting view of water cascading over the dam face (see sidebar).

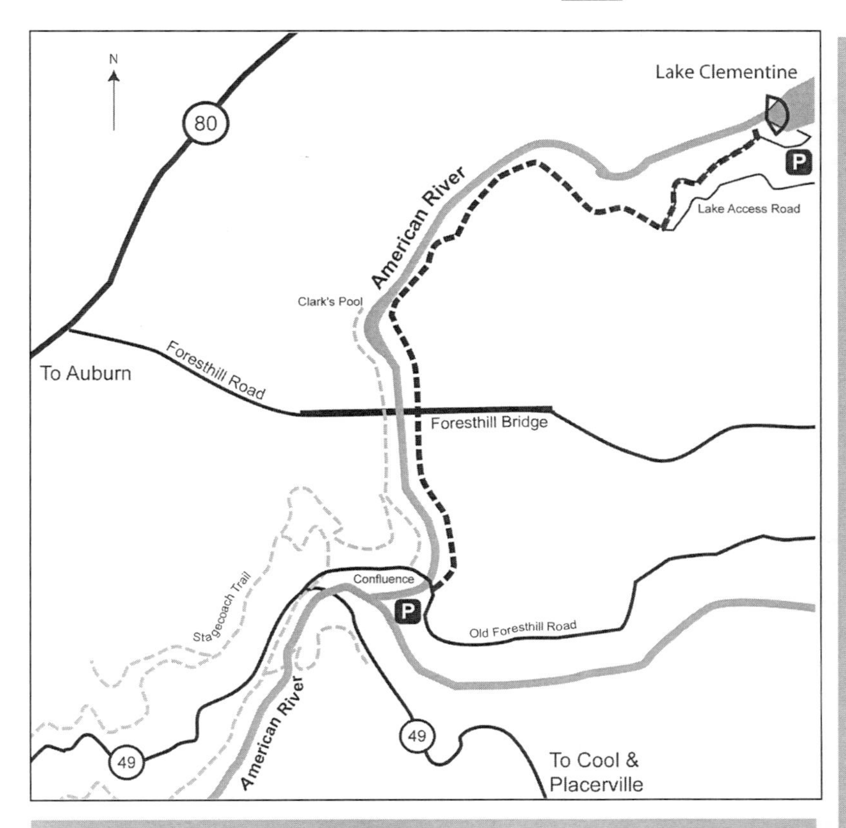

Did You Know? – The 2,248-ft long Foresthill Bridge was designed to span the reservoir that would have resulted had the Auburn Dam been completed. (Work on the dam was discontinued in 1976.) Water was expected to reach near the top of the cement piers. Today the bridge towers 730 feet above the river, making it the tallest bridge in California. It was opened in 1973 with much fanfare and has been featured in numerous movies and commercials. It has also been the site for many stunts – both legal and illegal.

Did You Know? -- Lake Clementine is used exclusively for public recreation. It was created in 1939 when the North Fork Dam was completed. The dam was designed and built by the Army Corps of Engineers in order to collect sedimentation from upriver hydraulic mining. A few years later, it was made superfluous when such operations were discontinued by state edict.

Did You Know? – Stagecoach Trail was originally a toll road built in 1852 known as Yankee Jim's Turnpike and later as Old Stagecoach Road. The original road crossed the North Fork American River at a toll bridge just upriver from Clarks Hole, and from there, it led to the towns of Yankee Jim's and Iowa Hill. In the late1800s, Yankee Jim's was a popular mining area, and Foresthill was yet to be developed. In 1875, the original toll bridge was replaced with a wooden covered bridge. In the 1870's, tolls on the bridge ranged from 6¢ for a cow to 50¢ for a horseman and $1 for a wagon and two horses.

Above: A short side trail leads hikers close to the dam face.

Left: Lake Clementine from the air.

Lakeview Connector Trail

Distance: 3.6 miles; 1½ hrs each way (hiking)

Difficulty: Easy to Moderate

Elevation Change: +/- 250 ft. (see below)

Trailhead/Parking: (N38-55-990; W121-00-705)

Trailhead is on Lake Clementine Rd. From I-80, take Foresthill Exit and continue on Foresthill Rd towards Foresthill for 3.2 miles; turn left on Lake Clementine Rd. A tiny parking area is ½ mile ahead on left at a sharp right bend in the road. **Watch for a dip when entering the parking area.** After parking, walk back along Lake Clementine Rd to trailhead you passed on the right just before the parking area. (An alternative parking area is ½ mile back up Lake Clementine Rd – you passed it on the right just after turning off Foresthill Rd.)

Description

This fairly level trail connects Lake Clementine Road to the Foresthill Divide Loop Trail, meandering through areas of chemise and foothill woodlands on the south side of the canyon above Lake Clementine. It affords some beautiful views of the lake, and it is one of the few places to see groupings of both madrone and manzanita trees. Caution: poison oak is plentiful along this trail.

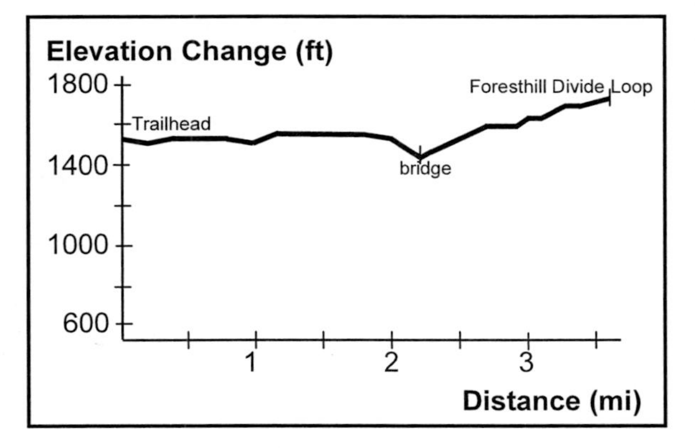

The Lakeview Connector Trail follows contours on the canyon between Lake Clementine (see sidebar) and Foresthill Road. On clear days, the vistas on this trail are outstanding. Although not visible until well along the trail, the lake and the opposite canyon occasionally peek through the trees at various points.

At the ¾ mile point, the trail opens up to views of beautiful, chemise-covered hillsides. A little further along the trail, large boulders appear to be strategically placed beside the trail as if a sculpture for your enjoyment.

At the 1 mile point, the trail crosses the unmarked but wide Lake Clementine Access Trail (N38-55-994; W121-00-692 - see separate trail guide) leading down to the lake in 1 ½ miles from the green gate (#115) on the Lake Clementine Rd and the back-up parking area.

The Lakeview Connector Trail continues straight across this trail, passing mainly through foothill woodlands for the next mile. There are several stands of manzanita and the less frequently seen madrone - both of which have distinctive reddish bark that always feels cool to the touch - as well as ponderosa pine, bay, several types of oaks and other foothill favorites.

At the 2¼-mile point, there is a bridge spanning a seasonal stream and its small riparian corridor. This is the lowest elevation on the trail. The trail then begins a steep rise for the next ½ mile, at the top of which is the best vista point on the trail, providing a great view of the lake. On the opposite hillside, below a few houses, a large outcropping of limestone can be seen. This is labeled "Lime Rock" on topo maps but is known locally as "Robbers Roost" (see sidebar). Look for the bench nestled among the boulders facing upstream for a perfect break spot.

Looking down on Lake Clementine from above it is clear that the lake is a dammed river, meandering through the North Fork American

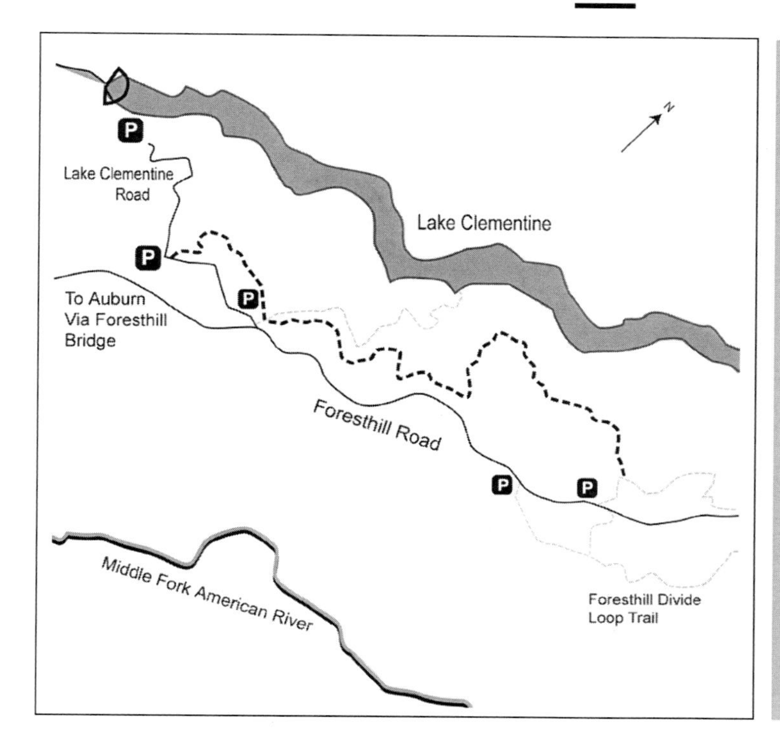

Did You Know? -- Lake Clementine is used exclusively for public recreation. It was created when the North Fork Dam was completed in 1939. The dam was designed and built by the Army Corps of Engineers in order to collect sedimentation from upriver hydraulic mining. A few years later, it was made superfluous when such operations were discontinued by state edict.

Did You Know? -- Lime Rock is locally known as Robbers Roost. When Foresthill was still a booming gold mining site, a local band of outlaws, the Gassaway Gang, is reported to have used the rock as a strategic lookout point to spot and then signal when the stagecoach was enroute from Foresthill to Auburn.

River canyon behind the spillway dam. a few miles downstream

The next mile is mostly in the shade, although small, open, sunny areas often have colorful displays of wildflowers in the spring. As you approach the end of the trail, there is a small clearing and a path entering from the left. This is a cutoff to the Foresthill Divide Loop Trail going towards Driver's Flat.

A little further along, there is a larger clearing and the Lakeview Connector Trail ends at the Foresthill Divide Loop Trail going both left and right. (See separate trail guide.)

Going right on the Foresthill Divide Loop, the trail passes alongside a pretty oak meadow and arrives at the Foresthill Road in less than 0.3 mile. There is a small dirt parking area on Foresthill Road, which some use as a shuttle parking area. The Loop Trail continues across Foresthill Road for another 8 miles, eventually returning to this intersection with the Lakeview Connector Trail.

Starting point of the Lakeview Connector Trail.

Lime Rock or Robbers Roost is one of the landmarks above Lake Clementine

Olmstead Loop Trail (#26 on the ASRA Topo Trail Map)

Distance: 8.6 miles; 4 hours (hiking), but a variety of shorter trail options/ cutoffs are also possible

Difficulty: Easy to moderate

Elevation Change: +/- 370 ft. (see below)

Trailhead/Parking: (N38-53-335; W121-01-036)

Trailhead and parking are behind the fire station in Cool. Take Hwy 49 south to Cool and turn right on St Florien Ct, before the fire station and blinking light. This is also the Cool staging area for equestrians. Trailhead is behind the wire fence at the south end of the equestrian parking area. The opening is at end of the fence nearest the stores and play yard.

Description

This beautiful loop trail, which parallels Hwy 49 on one side and the American River Canyon on the other, passes through open, rolling hills with several species of oak trees, and wildflowers in spring. It also includes steep canyon descents and climbs as it crosses Knickerbocker and Salt Creeks. Side trails on the canyon side offer panoramic views of the North Fork American River and of the inactive Auburn dam site. Trail markers have been posted about every half-mile and at most trail intersections.

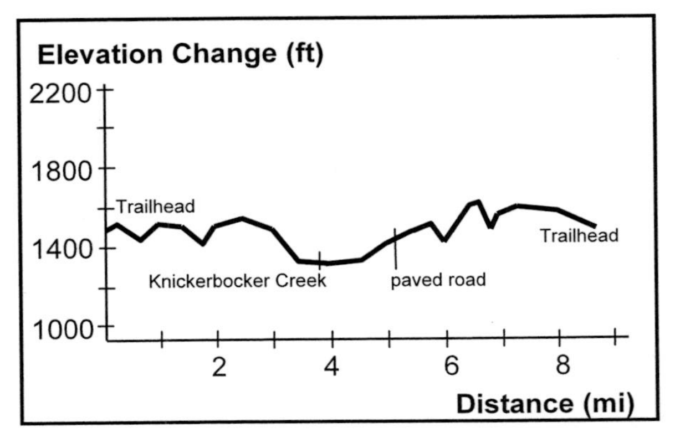

The Olmstead Loop Trail is especially popular in the springtime when wildflowers are in bloom, ponds and vernal pools are visible, and over 50 species of birds may be seen. Weekends attract many bikers and equestrians. Although bikers seem to be attracted to the mud, hikers are cautioned that after a heavy rain the trail may be very muddy since most of the trail holds water. Many unmarked paths crisscross the area. The main Olmstead Loop is always the widest trail, with green mile markers.

Proceeding around the loop in a clockwise direction, most of the first section of the trail is fairly flat and meanders through a typical foothill oak woodland ecosystem. Once through the fence, go straight. Soon the trail passes a small horse ranch surrounded by sprawling hills, providing great views of the snow-capped Sierras in the distance. At about 3½ miles, the trail starts winding down through a pine forest to Knickerbocker Creek. At the creek crossing, enjoy the natural setting with its pools and rushing water before heading uphill. The adventurous may want to scramble down the rocks to see a waterfall, pools, and riparian ecosystem along Knickerbocker Canyon.

Continuing on the Olmstead Loop, the trail is more arduous beyond Knickerbocker Creek, but after another half mile you will be at the top of a hill that affords a grand view. The trail soon crosses the wide, paved road that was built to provide access to the top of the keyway for the planned Auburn dam (to the left). For a shortcut, take this road to the right for 1¾ miles to get back to the Cool fire station (for about a 7-mile loop).

Crossing the paved road the Olmstead Loop Trail continues on for about another 4 miles. Soon after crossing the road, there is a trail inter-section, with one trail going uphill on the left, a cutoff from the road entering on the right, and the Salt Creek Loop Trail (see separate trail guide) starting downhill on the left. Continue straight up the hill.

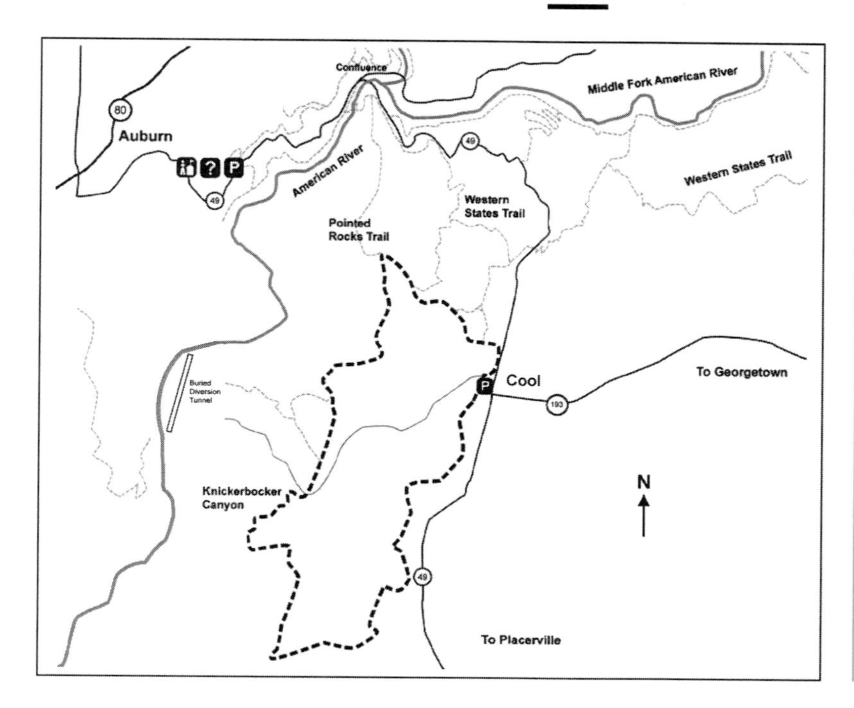

Did You Know? – The Olmstead Loop Trail was formerly known as the Knickerbocker Trail. It was named for Dan Olmstead in 1993. Olmstead was a local avid mountain biker and hiker. He dedicated himself to creating harmony between hikers, mountain bikers and equestrians, and to multi-use of trails in this area. He also established a bike and hiking shop in the area.

The Salt Creek Loop Trail is marked as the Auburn to Cool Trail and/ or Coffer Dam Trail on older signs. The next intersection with it will be shortly before a steep decent to Salt Creek. Past the creek crossing, a gradual climb affords lovely views. The terrain flattens out, and then another steady climb takes you to the top of a shady knoll. Here, the trail meets the Pointed Rocks Trail (see separate trail guide) which descends left to the confluence area.

The Olmstead Loop descends gradually after this, winding to the right in a big U-turn past out-croppings of pointed rocks. It passes two side trails that lead to the Western States Trail (labeled Wendell T. Robie Trail on the markers), which both take off to the left. Tevis Cup riders and 100-mile endurance runners use the WST for their annual Squaw Valley-to-Auburn races.

At the bottom of a hill, the Olmstead Loop crosses another small knoll before returning to the Cool fire station.

One of the many meadows traversed by the Olmstead Loop Trail.

Park HQ-to-Confluence Loop

Trails Included: Park Access Trail (#28)
Western States Trail (#42)
Tinker's Cut-off Trail (#38)
Stagecoach Trail (#36)
Manzanita Trail (#22)

Distance: 3½ mi. / 2 hrs. (hiking)

Difficulty: Easy to moderate

Elevation Change: +/- 450 ft. (see below)

Trailhead/Parking: (N38-54-300; W121-02-380)

Trailhead is on Hwy 49, opposite ASRA Park HQ, ¾ mi. south of Auburn behind the green gate (#136). Parking is at ASRA Park HQ. Alternate trailhead is at the confluence area, 1½ mi south of Park HQ. Take Hwy 49 south to Old Foresthill Rd. at bottom of canyon; go straight and park on right just past intersection.

Description

This loop affords many beautiful views of the American River Canyon and the confluence area. It also includes some of the most historic trails and sites in the Auburn State Recreation Area (ASRA). Although the trails may be steep at times, it is mostly an easy loop through three of common American River Canyon ecosystems: foothill woodlands, riparian woodlands, and chaparral.

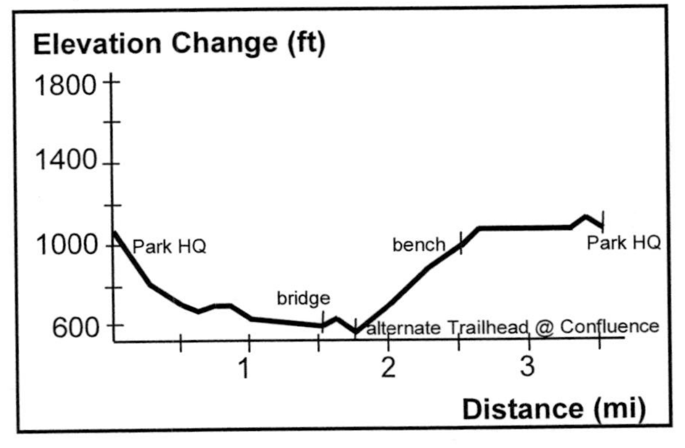

The loop starts at the ASRA Park Headquarters. Carefully cross Hwy 49 opposite the Park HQ and head down the Park Access Trail behind the metal green gate (#136). This trail is fairly steep, dropping 360 feet in less than a half mile. At the bottom of the hill, turn left onto the historic Western States Trail (see sidebar). This trail follows the old Mountain Quarries RR bed. Several old cement pilings for the railroad trusses can be seen along the trail. After a few minutes of hiking, you will come to a small stream crossing known as the "Black Hole of Calcutta" a wet, boulder-strewn stream crossing. There is a beautiful waterfall close to the trail, cascading over a natural rock wall.

The next mile is a gentle descent to the Mountain Quarries RR Bridge (see sidebar), with beautiful views of the American River along the way. Side trails lead down to sandy beaches and gentle rapids on the American River, popular with sunbathers on hot summer days. Approaching the Mt. Quarries RR Bridge, two other bridges can be seen in the distance. They are the Hwy 49 Bridge and the high Foresthill Bridge.

The loop continues via an unmarked path on your left just before the Mt. Quarries RR Bridge. This trail parallels Hwy 49 and goes under the Hwy 49 Bridge. Watch out for poison oak as it thrives here. This is a great place to view the confluence area (where the North and Middle Forks of the American Rivers converge). After reaching the road, cross to the other side and go up the steep hill just behind the green "Cool / Placerville" sign. This is the unmarked start of Tinker's Cut Off (see separate trail guide). This scenic trail is steep with many switchbacks, climbing 320 feet in just 0.3 mile. It passes through thick riparian woodlands as it parallels a small creek with several tiny waterfalls (best flows are in winter and spring).

At the top of Tinker's Cut Off, turn left onto the

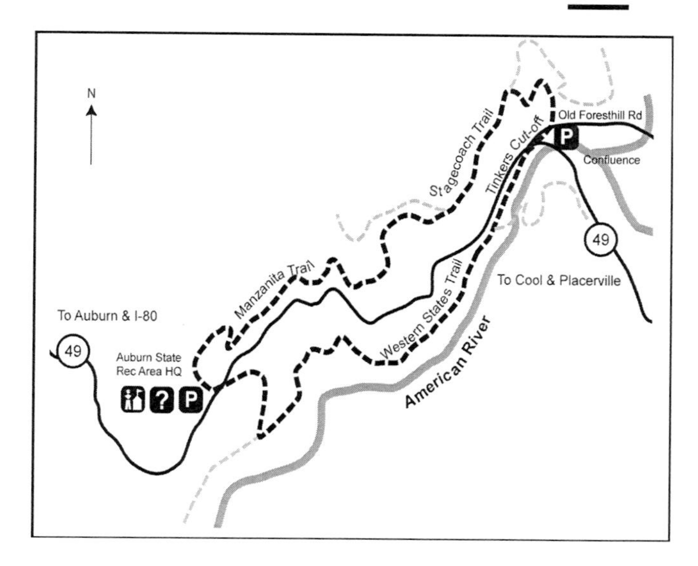

historic Stagecoach Trail (see separate trail guide). This is the actual road used by stagecoaches to carry passengers between Foresthill and Auburn in the late 1800s. Its wide, gentle gradient makes it a favorite of hikers and bikers throughout the year. Continue up the hill to a bench on the left, affording an opportunity to catch your breath, enjoy the view, and take more photos of the confluence area. As you sit, try to imagine you are sitting on the shore of a lake created by the proposed Auburn Dam (the high water mark of the lake would be just below your feet).

A few paces beyond the bench, go left on the unmarked Manzanita Trail, which intersects Stagecoach Trail at a sharp angle. Although narrower, Manzanita Trail is more level and meanders through chaparral where manzanita and coyote brush dominate the hillside. After a mile of fairly easy hiking, the trail ends at a firebreak road just above ASRA Park HQ. Look for white cable TV boxes and then take the firebreak road on the left past some buildings to return to Park HQ.

Did You Know? – The Western States Trail (WST) originally stretched from Sacramento to Utah. The Sierra Crest portion of the trail, blazed by Paiute and Washoe Indians and later used by miners, is now the route of two world-famous endurance races: *Tevis Cup Ride* for horses and *Western States 100-mile Endurance Run*. This part of the WST (from Mountain Quarries RR Bridge to the Auburn Staging Area) is the final leg of both of these endurance races.

Did You Know? – The Mountain Quarries Railroad Bridge was one of the world's largest reinforced concrete bridges when it was completed in 1912. It was used until 1940 by the Mt. Quarries Railroad to haul limestone from the nearby quarry up to the Southern Pacific RR main line in Auburn. The arched bridge has withstood several floods, including the disastrous one of 1964 resulting from the failure of Hell Hole Dam – a testament to its rugged design. In 1942, the railroad tracks were taken up for the war effort. Today, the bridge is the main artery for hikers and equestrians on the Western States Trail between Cool and Auburn. It was nicknamed "No Hands Bridge" by locals who liked to ride horses over it prior to the construction of the safety railing without holding onto the reins, shouting, "Look, no hands"! It was listed on the National Register of Historic Places in 2004.

A 1912 view of a portion of the Park HQ-to-Confluence Loop Trail as it approaches the Mountain Quarries RR Bridge.

PG&E Road Trail (#29 on the ASRA Topo Trail Map)

Distance: 3.6 miles; 2 hours one way (hiking) from top to Quarry lot.
Trail itself: 1.5miles
Overlooks: 0.6 miles
Quarry trail: 1.5 miles

Difficulty: Easy down, Moderate up

Elevation Change: +/- 700 ft. (see below)

Trailhead/Parking

North: (N38-54-713; W121-02-095)

South: (N38-54-515; W121-01-029)

Trailhead (north) is 2 miles south of ASRA Park Headquarters at the Quarry lot. Take Hwy 49 south from Auburn, turn right across the American River towards Cool. Turn left down a small dirt road ¼ mile south of the river crossing. Trailhead is behind green gate (#151).

Trailhead (south) is a tiny unmarked parking area for about 2 cars at the top of the trail off Hwy 49. At 1.9 miles after crossing the Hwy 49 Bridge to Cool, look for the large boulders on the left. It is best to go another 0.4m and turn around at the business entrance to the active Cool Cavern Quarry. This makes entry into the turn out easier and safer. A shuttle works well.

Description

This dramatic and scenic hike offers spectacular views of the Middle Fork as well as the results of present and past quarrying operations. Limestone has been extracted from this site since the 1880's. The current quarrying operations are limited to the upper ridge. The trail is accessed either from the Quarry Trail (see separate trail guide) or from a small turn out off Hwy 49. Due to limited parking at the top, it is best to shuttle to the top and hike back to the quarry parking lot. Take time to visit the two short spurs to enjoy the views upriver as well the unique views of the dramatic rock remains of earlier quarrying operations.

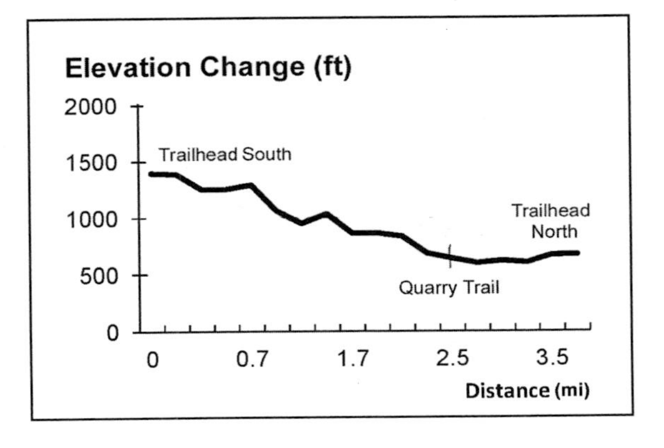

The PG&E Trail from the top begins behind the large boulders marking the small pull out off Hwy 49. The trail immediately crosses under one of the PG&E power poles for which it is named. The upper section is narrow, as it winds its way through shrubbery and poison oak. The current quarry operation, on the right, may be heard but is screened by a fence and foliage. The trail gently works its way downhill. After passing another power pole, a transformer may be seen and heard on the left before coming to an opening where the main gravel access road for PG&E enters from the left. Go right and continue down the trail.

As the trail bends to the left and becomes wider and steeper, detour to the right along the narrow trail visible here for about 0.1 miles to the first overlook. Through the trees on the left, there are nice views up the Middle Fork. You will pass an old building left from when this was an active quarry site. The trail ends at a wide open bench area. You can often view machinery working on current quarrying activities in the distance as well as viewing the rocky sentinels left by past mining activities. It is fun to watch the vultures circling below you. Return to the main trail which is now wider and steeper.

Next the trail intercepts another old mining road. Again, the main trail goes left, but a short detour to the right is well worth the time. You will pass two abandoned mining buildings on this old road

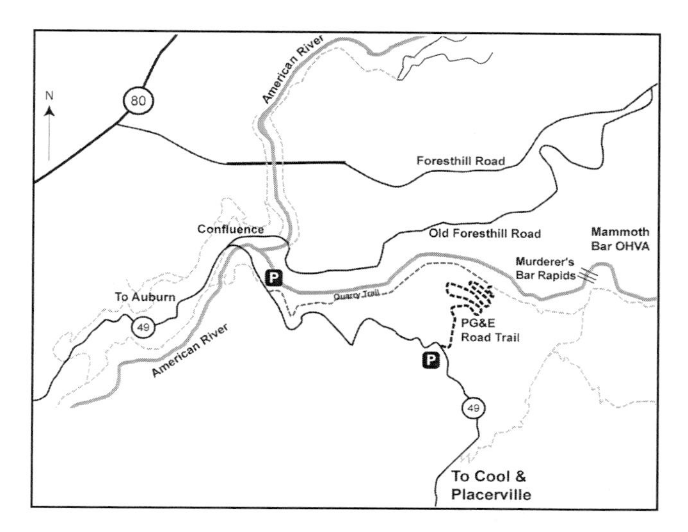

before coming to a large open area. This open section affords great views up the Middle Fork, down on the valley created by the earlier extraction of limestone and alongside and above to the scarred cliffs remaining from the extraction process. The trail terminates at a fence making the boundary of the ASRA.

After taking in the views, return to the main trail as it winds its way via a series of switchbacks down towards the Quarry Trail. When you reach a road going left and right, the main trail continues to the left but go right to enter an older section of quarry – known locally as the amphitheater. This is a popular break area for users of the Quarry Trail. After entering the amphitheater clearing you can look up and see the last bench you visited. From this special place, return and continue down the PG&E trail. In a few turns you will intercept the Quarry Trail. Go left down towards the river. Continue downstream on the level Quarry Trail for approximately 1.2 miles to the Quarry parking lot.

Did You Know? – The Quarry was first known as the Cool Cave Quarry. In the 1860's, Munson Manning started a lime kiln here. In 1910, the Mountain Quarries Co. started work on the limestone quarry and railroad to carry its products to the main rail line in Auburn. Both went into operation in March 1912. The Bureau of Reclamation bought the right of way and underground site of the Mountain Quarries when work began on the Auburn Dam project. Spreckles Sugar Company operated the upper quarry from the 1940's until 1987. Teichert is the current operator.

The quarry contains the largest exposed limestone deposit in the Sierra Nevada. It is very pure with a rating of 96-98%. From 1912 to1930, the quarry was one of the chief sources of limestone for the cement, sugar and steel industries in northern California.

Most of the limestone consists of ancient marine sediments from the Paleozoic era that were uplifted and folded into what is known as the Calaveras Formation. The rock is a part of a ridge of limestone running from Calaveras to Lake Oroville. Across the river you can see another outcropping.

Looking up the Middle Fork American River.

Rock sentinels from past mining.

Pointed Rocks Trail (#30 on the ASRA Topo Trail Map)

Distance: 1.6 miles one way; 1½ hrs up, ¾ hrs down (hiking). This does not include distance to trailhead. A variety of loop options are available.

Difficulty: Difficult to moderate

Elevation Change: +/- 1,050 ft. (see below)

Trailhead/Parking: (N38-54-893; W121-02-388)

Trailhead is on Hwy 49, 1¾ miles south of ASRA Park Headquarters. Take Hwy 49 south from Auburn towards Placerville. After crossing the American River, park on the right off the highway. Walk to trailhead through the green gate (#150).

Description

This steep trail offers great bird's eye views of the confluence and American River canyons. On clear days, it offers sweeping views of the Sacramento Valley to the west and the Sierra Nevada to the east. Its steep gradient offers a good aerobic workout, climbing 1000 ft in 1.2 miles. It follows a buried telephone cable on its ascent from the confluence area to the ridge and the Olmstead Loop. After reaching the ridge, the trail passes through open, rolling hills and meadows dominated by several species of oaks, and wildflowers in the spring. Limestone out-croppings at the top give the trail its name.

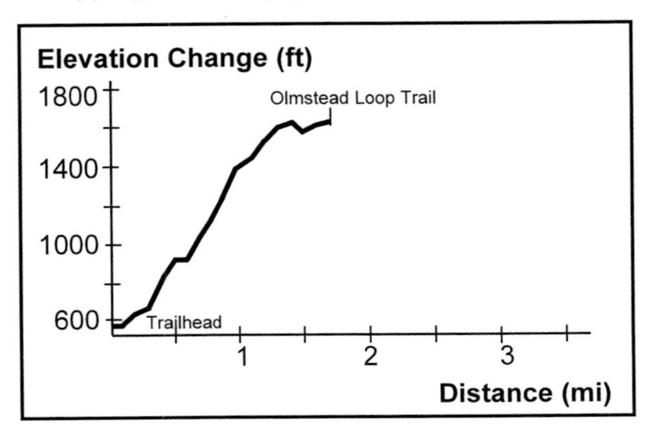

The Pointed Rocks Trail is sometimes referred to as "The Training Hill" on trail markers since it offers a great workout for endurance runners and equestrians. It provides one of the steepest gradients of any trail in the area. (Trail markers are posted at some, but not all intersections.)

Walk downstream on the trail behind the green gate on Highway 49 on the south side of the bridge over the American River. This wide road parallels the river down-stream for about 0.1 mile towards the Mountain Quarries RR Bridge (see sidebar). You may wish to cross this graceful arched bridge spanning the American River to read the plaque on the far side that describes its history.

Just before reaching the Mt. Quarries RR Bridge, a narrow trail goes up the hillside on the left (look for a sign to Cool - N38-53-935; W121-01-996). This is the start of both the Pointed Rocks Trail and the Western States Trail (WST). (see sidebar). The trail immediately switchbacks and begins climbing, offering some nice over-head views of the Mt. Quarries RR Bridge.

In less than 0.2 miles, the trail splits. The Pointed Rocks Trail veers up to the right while the WST continues around the ridge. Take the unsigned Pointed Rocks Trail to the right. The trail soon widens and starts to climb sharply.

In less than ½ mile, nice views of the American River and Hwy 49 as it wends its way up to Auburn can be seen through openings in the ponderosa, oak, manzanita, buck brush and various other foothill trees that line the trail. Near the top of the ridge, look back and east for a view of the Sierra Nevada and the Middle Fork American River.

After reaching the ridge, the terrain flattens out and meanders through a picturesque rolling oak woodland. Watch for fine-grained sandstone outcroppings of pointed rocks for which the trail is named.

After reaching an oak meadow, a small footpath

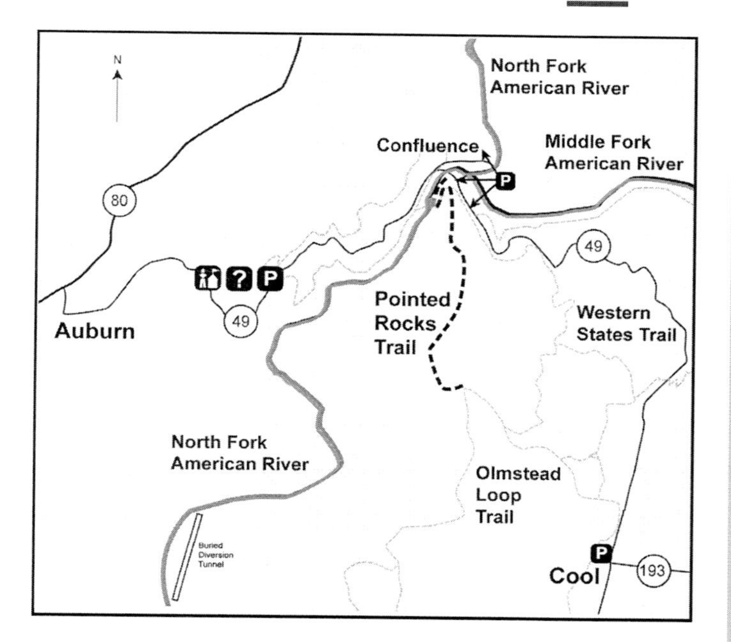

splits off to the left. The main, wider trail continues around this meadow, offering a panoramic view on clear days of Folsom Lake and Sacramento. Both paths soon rejoin.

A little further along, the Pointed Rocks Trail ends at the Olmstead Loop Trail. There are several options at this point: a) turn around and return to the confluence the same way, b) make a loop using the WST for a gentler but longer return trip, or c) hike into Cool using the Olmstead Loop. The shortest way to Cool is to continue straight on the Olmstead Loop for another 1.7 miles. (See separate guide)

The WST branches off the Olmstead Loop a short distance ahead on the left. It is labeled Wendell T. Robie Trail on some trail markers. The 100-mile endurance runners and riders both use this section of the WST for their annual Squaw Valley to Auburn races (see sidebar).

Did You Know? – The Mountain Quarries Railroad Bridge was one of the world's largest reinforced concrete bridges when it was completed in 1912. It was used for about 30 years by the Mt. Quarries Railroad to haul limestone from the nearby quarry up to the Southern Pacific RR main line in Auburn. The arched bridge has withstood several floods, including the disastrous one of 1964 resulting from the failure of Hell Hole Dam – a testament to its rugged design. In 1942, the railroad tracks were taken up for the war effort. Today, the bridge is the main artery for hikers and equestrians on the Western States Trail between Cool and Auburn. It was nick-named "No Hands Bridge" by locals who liked to ride horses over it prior to the construction of the safety railing without holding onto the reins, shouting, "Look, no hands"! It was listed on the National Register of Historic Places in 2004.

Did You Know? – The Western States Trail (WST) originally stretched from Sacramento to Utah. The Sierra Crest portion of the trail, blazed by Paiute and Washoe Indians and later used by miners, is now the route of two world-famous endurance races: *Tevis Cup Ride* for horses and *Western States 100-mile Endurance Run*.

Some of the rocks that gave the Pointed Rocks Trail its name.

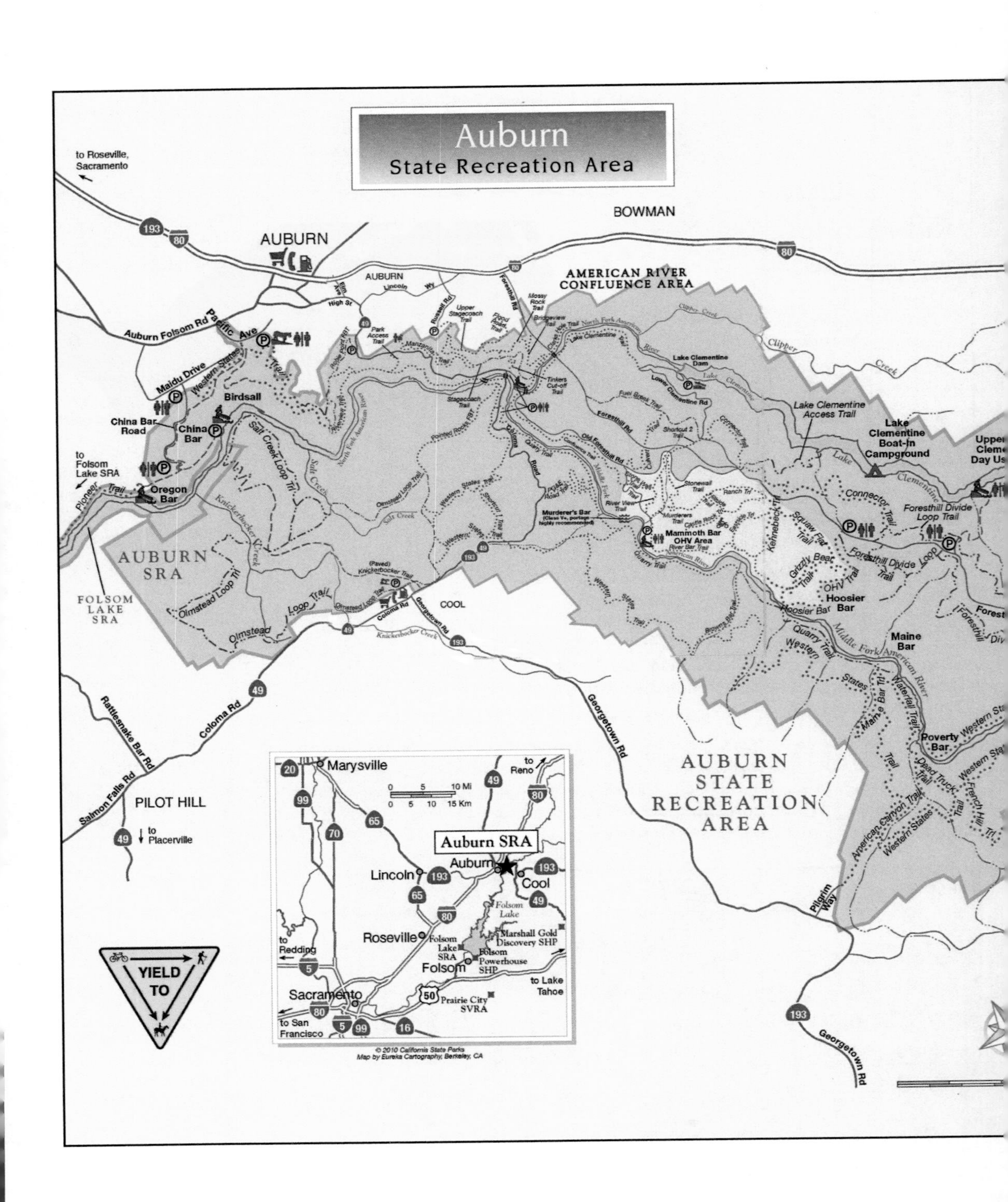
Auburn
State Recreation Area
to Roseville, Sacramento
BOWMAN
AUBURN
AMERICAN RIVER CONFLUENCE AREA
Auburn Folsom Rd
Pacific Ave
Maidu Drive
Birdsall
China Bar Road
China Bar
to Folsom Lake SRA
Oregon Bar
Pioneer Trail
Salt Creek Loop Trl
Knickerbocker Creek
AUBURN SRA
FOLSOM LAKE SRA
Olmstead Loop Trl
Olmstead Loop Trail
Western States Trail
Murderer's Bar
Mammoth Bar OHV Area
Lake Clementine Dam
Lower Clementine Rd
Foresthill Rd
Old Foresthill Rd
Lake Clementine Access Trail
Lake Clementine Boat-In Campground
Connector Trail
Foresthill Divide Loop Trail
Clipper Creek
COOL
Coloma Rd
Georgetown Rd
Hoosier Bar
Maine Bar
Poverty Bar
Quarry Trail
Middle Fork American River
AUBURN STATE RECREATION AREA
Rattlesnake Bar Rd
Salmon Falls Rd
PILOT HILL
to Placerville
Pilgrim Way
American Canyon Trail
Dead Truck Trail
YIELD TO
Marysville
to Reno
Auburn SRA
Auburn
Lincoln
Cool
Folsom Lake
Roseville
Folsom Lake SRA
Marshall Gold Discovery SHP
Folsom
Folsom Powerhouse SHP
to Redding
Sacramento
to Lake Tahoe
Prairie City SVRA
to San Francisco
© 2010 California State Parks
Map by Eureka Cartography, Berkeley, CA

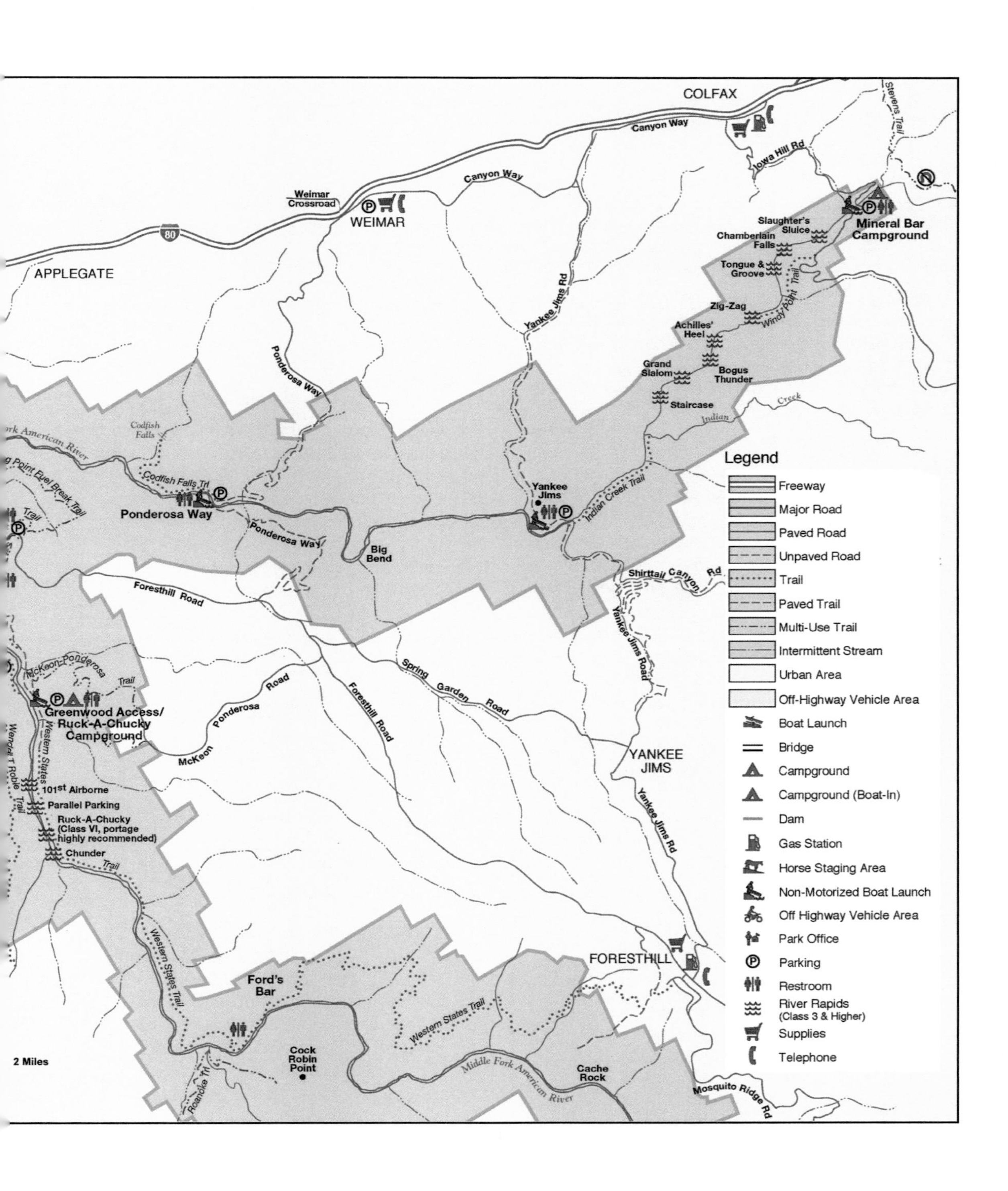

COLFAX
Canyon Way
Iowa Hill Rd
Stevens Trail
Weimar Crossroad
WEIMAR
Canyon Way
APPLEGATE
80
Mineral Bar Campground
Slaughter's Sluice
Chamberlain Falls
Tongue & Groove
Windy Point Trail
Zig-Zag
Achilles' Heel
Grand Slalom
Bogus Thunder
Staircase
Indian Creek
Yankee Jims Rd
Ponderosa Way
Codfish Falls
Codfish Falls Trl
Ponderosa Way
Ponderosa Way
Big Bend
Yankee Jims
Indian Creek Trail
Shirttail Canyon Rd
Foresthill Road
Yankee Jims Road
Spring Garden Road
Foresthill Road
McKeon-Ponderosa Trail
Greenwood Access/ Ruck-A-Chucky Campground
McKeon Ponderosa Road
YANKEE JIMS
Yankee Jims Rd
Western States
Wendell T Robie Trail
101st Airborne
Parallel Parking
Ruck-A-Chucky (Class VI, portage highly recommended)
Chunder
Trail
Western States Trail
FORESTHILL
Ford's Bar
Western States Trail
Cock Robin Point
Roanoke Trl
Middle Fork American River
Cache Rock
Mosquito Ridge Rd
2 Miles
Legend
Freeway
Major Road
Paved Road
Unpaved Road
Trail
Paved Trail
Multi-Use Trail
Intermittent Stream
Urban Area
Off-Highway Vehicle Area
Boat Launch
Bridge
Campground
Campground (Boat-In)
Dam
Gas Station
Horse Staging Area
Non-Motorized Boat Launch
Off Highway Vehicle Area
Park Office
Parking
Restroom
River Rapids (Class 3 & Higher)
Supplies
Telephone

Quarry Road Trail (#31 on the ASRA Topo Trail Map)

Distance: 5.6 miles; 2¼ hours each way (hiking)

Difficulty: Easy

Elevation Change: +/- 210 ft. (see below)

Trailhead / Parking: (N38-54-713; W121-02-095)

Trailhead is 2 miles south of ASRA Park Headquarters. Take Hwy 49 south from Auburn, turn right across the American River towards Cool. Turn left on a small dirt road ¼ mile south of the river crossing. Trailhead is beyond the parking area at a green gate (#151).

Description

This wide, level trail can be used for a half-day walk and a pleasant picnic along the Middle Fork (MF) American River. It goes through some of the best scenery available in the American River Canyons, and picnic tables are provided along the first 1¼ miles of the trail. For those looking for a good workout, this trail is 11.2 miles round-trip. Several side trails are accessible that can be used to create even more challenging loops. There is little shade, however, so take plenty of water and sunscreen on hot summer days.

The Quarry Road Trail follows the route of the gold rush era Grand Flume, a 13-mile long wood and canvas flume that was built annually by private mining companies to extract gold from the river during the late 1850's. The Mountain Quarries Railroad also used the first 1¼ miles of the trail in the early 1900's to transport limestone from the quarry up to Auburn.

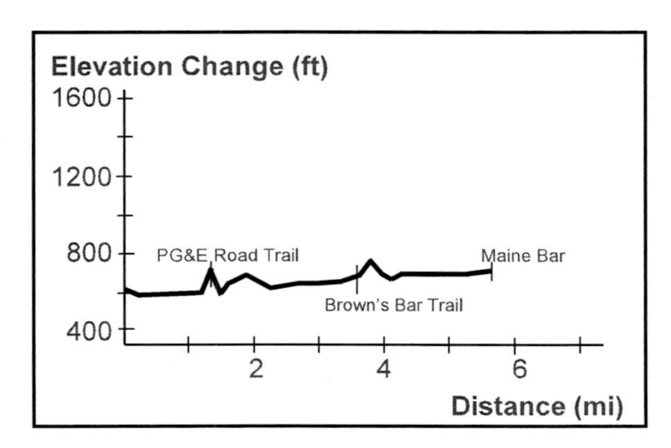

After 1¼ miles, the trail bears right up a short hill above the picnic area. Remnants of a large limestone-loading platform for the rail cars can still be seen on the left. At the top of the hill, the unmarked PG&E Road Trail (see separate trail guide) takes off uphill from just above the ruins of the limestone-loading platform.

The PG&E Road Trail (hikers and bikers only- see separate guide) offers fascinating views into the now abandoned quarry. A short uphill trek and a couple of switchbacks bring you to a spectacular amphitheater in the older section of the quarry – well worth a 20-minute detour. For those with enough stamina to take this 1.5 mile trail to the top, it offers some awesome viewing spots of the river canyon and quarry. It comes out on Hwy 49 at the top.

Returning to the Quarry Road Trail and continuing upriver, the trail meets the historic Western States Trail (WST) at the 2-mile point. In the river below, you can hear the Murderer's Bar Rapids, named for a deadly skirmish between miners and Native Americans that occurred here in 1849. Further upriver, you may see or hear off-highway vehicles scurrying up and down the canyon on the other side of the river in the Mammoth Bar Off-Highway Vehicle Area.

At the 3.5-mile point, the trail intersects with Brown's Bar Trail, which heads uphill to the WST. The Quarry Trail continues parallel to the MF American River for another 2 miles. Just before reaching Maine Bar, the trail intersects with Maine Bar Trail. Like the earlier Brown's Bar Trail, this too heads uphill along a creek and intersects at the top with the higher WST. No bikers are permitted on any of these side trails. The Quarry Road Trail ends at Maine Bar.

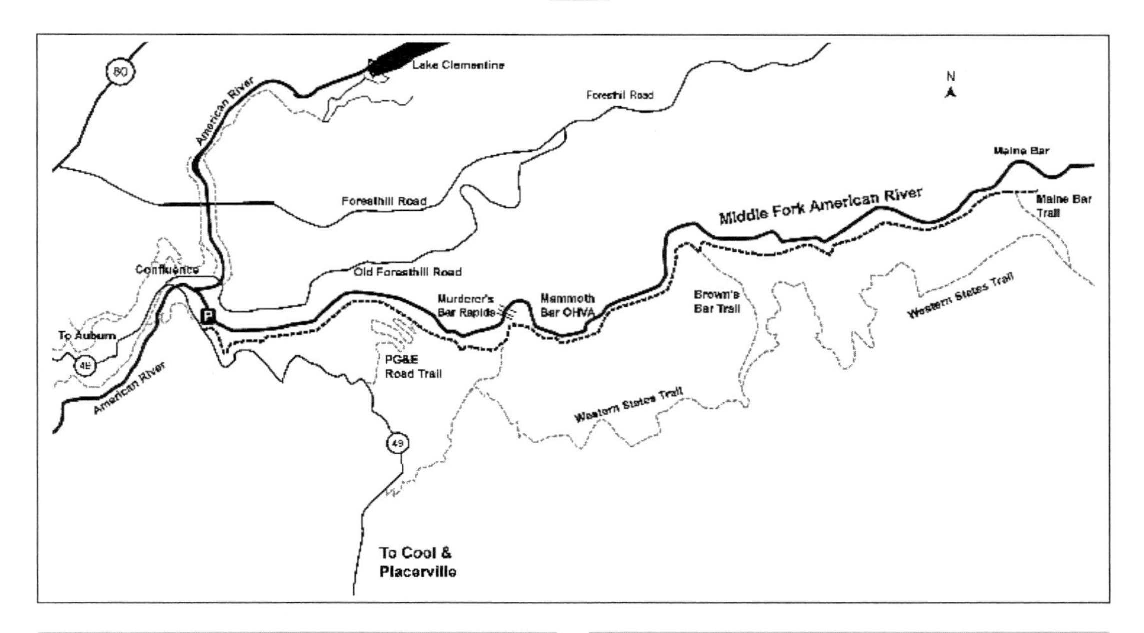

Did you know? – The gated entrance to the Mountain Quarries Mine, that contains some remnants of Hawver Cave, is readily visible just off the Quarry Trail. It is just beyond the tall concrete remains, about 1¼ mile from the trailhead.

The cave was named after Auburn dentist and amateur paleontologist Dr. J. Hawver, who brought the cave's importance to light. Between 1906 and 1910, Dr. Hawver with the assistance of UC archaeologists, found and preserved many remains from the cave.

These included 10,000-year-old human bones and many fossils including those of the dire wolf, saber-toothed cat, a giant ground sloth and other prehistoric animals. The limestone mining operation, which started in 1911, eventually destroyed virtually all of Hawver Cave, but some small remnants of natural cave features remain within the mine.

There are two local displays featuring Hawver Cave, one at the Placer County Museum and the other at the Sierra College Library.

The Quarry Trail today, looking downstream towards the trailhead.

Looking upstream at portions of the old quarry and Quarry Trail in the early 1930's.

Quarry & Western States Loop Trail

Trails Included: Quarry Trail (#31)
Western States Trail Tevis (#44)
Western States Trail (#43)
Shortcut Trail 1 (optional) (#34)

Distance: 6.3 mi. / 3 hrs. (hiking)

Difficulty: Moderate to Difficult

Elevation Change: +/- 1,000 ft. (see below)

Trailhead/Parking: (N38-54-713; W121-02-095)

Trailhead is 2 miles south of ASRA Park Headquarters. Take Hwy 49 south from Auburn; turn right across the American River towards Cool. Turn left on a small road ¼ mile south of the river crossing. Trailhead is beyond the parking area at green gate (#151).

Description

The Quarry Trail portion of this loop follows the wide, level bed of the Mt Quarries RR. It affords nice views of the MF American River. The two sections of the Western States Trail (WST) cut through the dense foothill woodlands above the river. This loop affords regular users of the Quarry Trail an alternate route rather than the normal out and back. It highlights a small portion of the WST while skirting the quarry operation. The loop goes from river to canyon rim and back.

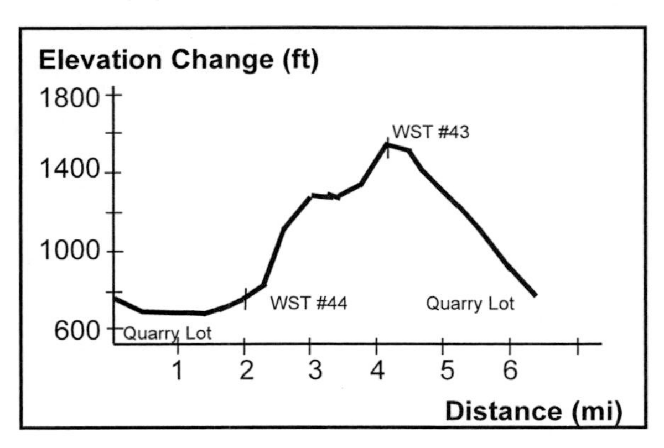

The loop starts at the Quarry parking lot, behind green gate #151. For the first 1¼ mile it follows the roadbed used by the Mountain Quarries Railroad to carry limestone to Auburn. Just before a large open picnic area, the trail bears right up a short hill, past remnants of the large limestone-loading platform for the rail cars.

At the top of the hill, a separate unmarked trail (PG&E Trail for hikers and bikers-see separate guide) goes uphill just above the ruins of the limestone-loading platform. A short trek via a few switchbacks leads to an impressive amphitheater in an older section of the quarry.

The main Quarry Trail continues straight past the gated Hawver Cave. The cave, the initial quarry site (see sidebar) contained many bones from the Pleistocene era. Continuing up river, the trail crosses a small stream before meeting the Western States Trail Tevis (WST–see sidebar) on the right at the 2-mile point. (N 38-55-018; W121-00-344). Turn right at the sign to Cool and WST, onto the narrow, steep but shaded trail going uphill. The occasional mile markers indicate the distance to the finish line in Auburn for Tevis and WST participants. This section may be slippery after rain. Watch out for poison oak on the remainder of the loop.

This portion of the WST climbs from the river to the rim through brushy undergrowth and foothills woodland with glimpses of the MF below and the limestone operations along the canyon rim. After climbing 0.6 of a mile, the loop intersects the main WST on the left. Continue uphill and in 0.1 mile the trail briefly uses a wide gravel road going into the active quarry site. Go left and look for the narrow dirt trail on the left just before the locked gate. There will be glimpses of the active limestone quarry on the right. Continue straight. Hwy 49 will soon be heard and then seen. There is a cross walk, with push button, across the loading and entrance area of the Teichert Cool Quarry. This section of the trail ends at Hwy 49 and Gate #153.

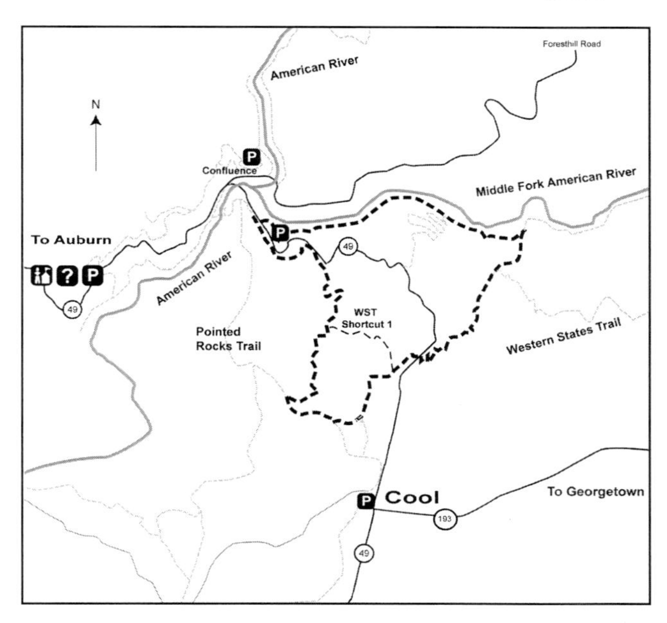

Carefully cross Hwy 49 to the clearing on the other side and go left. After 0.1 of a mile, the Shortcut Trail 1 will be on the right. Taking this will cut ½ mile off the loop. If you continue straight, going uphill towards Cool, the Paige Harper Spring watering trough, which captures the natural creek water for horses, will be on your right.

As you come to the open oak grasslands of the Olmstead Loop area, take the unmarked cut off to the right and then the next right, following the signs to the WST that are also signed as "Wendell T Robie" or "To Auburn". Enjoy the open vistas of the oak meadows before the WST again heads into the trees and shrubs on its way down to the river. The gentle downhill section generally parallels a small creek and Hwy 49. It offers dappled shade, a few small water crossings, and glimpses of Hwy 49 on the right. The Shortcut Trail 1 joins the WST at about the 5 mile point. At 5.8 miles, there is a steep uphill trail known as the "pig farm trail" that goes off on the left. It is mainly used by equestrians preparing for the Tevis Cup. Continuing downhill, the tall Foresthill Bridge and confluence area will soon appear in the openings. Take the next right (about 6.2 miles) down a short steep section to Hwy 49 directly across from the quarry trail parking area. Cross Hwy 49 carefully.

The loop may be done in reverse order, but since Tevis riders and those practicing for it usually go in the direction indicated, the grooves and ruts generally make it easier in this direction. In some of the steeper sections, repeated horse traffic has created a trench like trail.

Did You Know? – The Quarry was first known as the Cool Cave Quarry. In the 1860's Munson Manning started a limekiln here. In 1910, the Pacific Portland Cement Company began its Mountain Quarries at the site. It built a railroad and the Mt. Quarries RR Bridge to carry its products to the main rail line in Auburn. The Bureau of Reclamation bought the right of way and under-ground site of the Mountain Quarries when work began on the proposed Auburn Dam. Spreckles Sugar Company operated the quarry from the 1930's until 1987. Teichert is the current operator.

The quarry contains the largest exposed limestone deposit in the Sierra Nevada. This same limestone vein also juts up above ground on the NF, notably as Lime Rock or Robber's Roost. The limestone deposit is extremely pure with an average purity rating of 96-98% that historically led to its use in the food and beverage industry, as well as in construction.

The Mountain Quarries operation in the 1930's.

Robie Point FB Trail (#33 on the ASRA Topo Trail Map)

Distance: 3.6 miles; 1½ hours each way (hiking), but a variety of shorter options are available

Difficulty: Easy to Moderate

Elevation Change: +/- 250 ft. (see below)

Trailhead/Parking:

North: (N38-54-001; W121-03-536)

South: (N38-53-245; W121-04-246)

Trailhead (north) is on Hwy 49, southbound from Auburn ½ mi (on the right) or northbound from ASRA Park Headquarters ¼ mi (on the left), behind gate #130. **Caution: cars coming downhill are hard to see on the curve.**

Trailhead (south) is at Overlook Park behind the Gold Country Fairgrounds. Take Hwy 49 and Lincoln Way north through Auburn to Auburn-Folsom Road, then past the fairgrounds and turn left on Pacific Ave. Turn right after 0.5 mi into the large Overlook parking area. After parking, walk back along Pacific Ave to trailhead to the equestrian staging area at a yellow-gated area that was on left as you entered the parking area.

Description

This firebreak trail is wide and well maintained. It provides great views of the North Fork (NF) American River canyon and the Auburn Dam construction site. Although the trail is steep at the southern end, it is otherwise fairly gentle and passes through three of the most common American River canyon ecosystems:

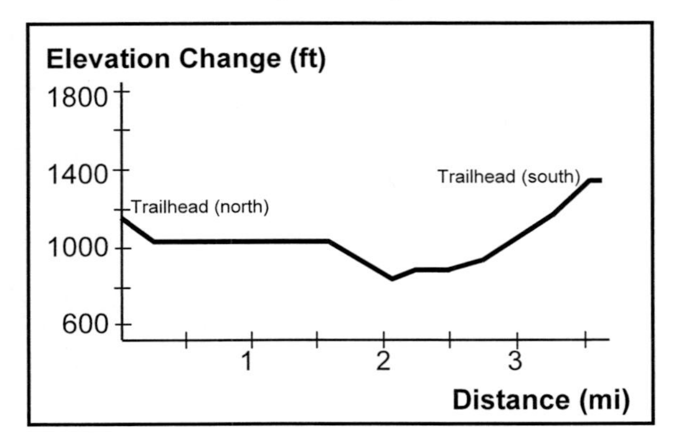

foothill woodlands, riparian woodlands, and chaparral. It can be hot in the summer since there is little shade, but it is ideal in the late afternoon or early evening for a hike or ride.

The Robie Point Firebreak Trail, starting at the north end (at gate #130), gradually descends to a small creek, passing through a densely forested oak and pine riparian woodland. The first 500 ft is paved, and then turns to packed dirt. After crossing the creek, the trail soon enters the main American River Canyon with some awesome views of the NF American River. Looking east, you can see the snow-capped Sierra Nevada in the distance.

At the 0.5 mile point, the trail merges with the historic Western States Trail (WST) (see sidebar), which enters from the left (a narrow dirt trail). At ¾ miles, the WST splits off to the right up the hill. For those wanting more of a workout, take the WST for ¼ mile up and over Robie Point. Otherwise, continue to the left on the Robie Point FBT for a level detour around Robie Point. Just past this point, you start to see the high Foresthill Bridge in the distance to the east.

Just beyond the 1-mile point, evidence of the Auburn Dam construction project becomes visible on the opposite side of the canyon. The terraced roadways up and down the canyon provided access for construction equipment to the keyways for the proposed dam. The trail rejoins the WST just past the 1¼-mile point. Stay to the left at the junction of the trails.

Just beyond the 1.5-mile point, the WST splits off (again) to the right up the hill around the gate. The Robie Point FBT continues to the left down the hill. From this vantage point, the work done on the hillsides for the proposed dam is easily visible. This was once the site of the now closed diversion tunnel (see sidebar).

From this viewpoint, the Robie Point FBT descends rapidly for the next ¼ mile. At the bottom of the hill, a side trail splits off to the left, leading down to the river and the boulder-strewn

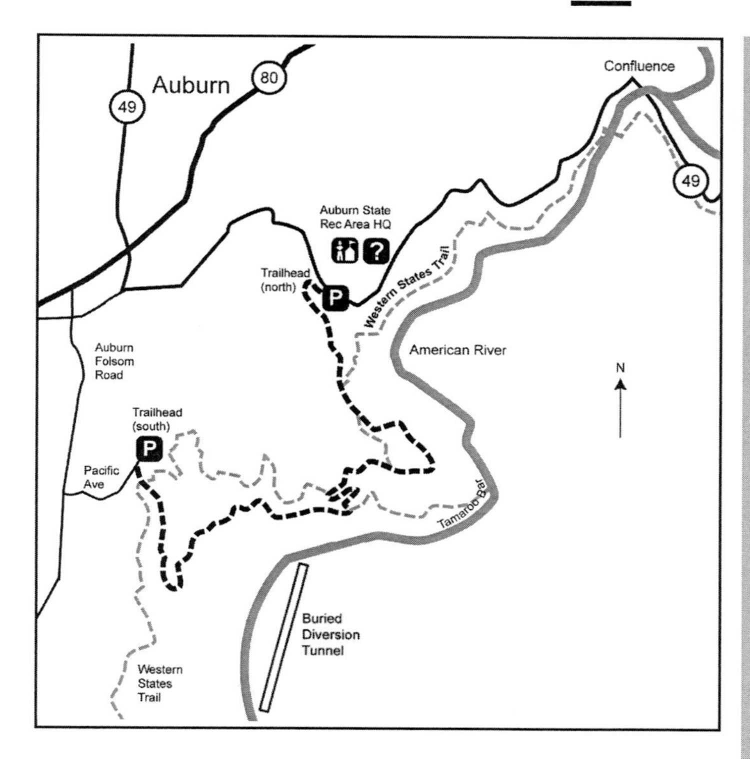

Did You Know? – The Western States Trail originally stretched from Sacramento to Utah. The Sierra Crest portion of the trail, blazed by Paiute and Washoe Indians and later used by miners, is now the route of two world-famous endurance races: the Western States 100 mile Endurance Run and the Tevis Cup Ride for equestrians. This part of the WST (from the Mountain Quarries RR Bridge to the Auburn Staging Area) is the final leg of these races. The trail marker at the base of Robie Point indicates "mile 2.0", which means there are two miles to the finish line.

Did You Know? – The diversion tunnel was 0.5 mile long and 30-ft in diameter. It was opened in 1968 to redirect the NF American River for the Auburn Dam construction. It was closed in 2008, and the river restored to its original channel where the man-made rapids and pump station are now located. The Auburn Dam was first designed as a double-curvature, thin-arch concrete dam. Extensive seismic studies of the dam site were completed in 1978, and as a result, the design was changed to a concrete gravity dam to address earthquake concerns. Work on the dam was later discontinued due to a lack of funding and other concerns.

Tamaroo Bar (¾ mile and a 400-ft descent, see separate guide). Robie Point FBT continues to the right for another 1.5 miles and affords even closer views of the dam construction site. The last ¾ mile of the trail climbs over 500 ft before reaching Pacific Ave.

If starting from Overlook Park, the trailhead (south) is on Pacific Ave at a paved and gated road just south of the parking area. Go around the gate and continue on the paved road down the hill. The road splits after 100 yards; follow the left fork down the hill. (The right fork is the former Auburn to Cool Trail.)

Left: View of the canyon.

Right: Hikers on the trail.

Salt Creek Loop Trail

Distance: 7.5 miles; 4 hours (hiking). .

Difficulty: Easy down, moderate to difficult up

Elevation Change: +/- 970 ft. (see below)

Trailhead/Parking: (N38-53-335; W121-01-036) Trailhead and parking are behind the fire station in Cool. Take Hwy 49 south to Cool and turn right on St Florian Ct, before the fire station and blinking light. Trailhead is on the paved road next to parking area behind Knickerbocker Gate #155. (N38-53-323; W121-01-120)

Description

The trail crosses the open, rolling oak studded hills of the Olmstead Loop. The descent down to the American River and back offers varied views of the river and the Auburn dam construction site. It provides a look at the scars on the canyon walls and rock debris left from the initial work done for the now dormant Auburn Dam project.

The Salt Creek Loop Trail offers a wide mix of scenery. It passes through open meadows, foothill woodlands, goes into the canyon, and passes along the American river and local creeks. The initial 1.3 miles is on the paved road built to access the top of the keyway for the proposed dam project. There are views of the area's oak woodlands and meadows. Knickerbocker Creek is visible on the left. Its course is frequently marked by the invasive Himalayan blackberry bushes that line its banks. The dirt trail that follows the creek is the Knickerbocker Trail and an optional return route.

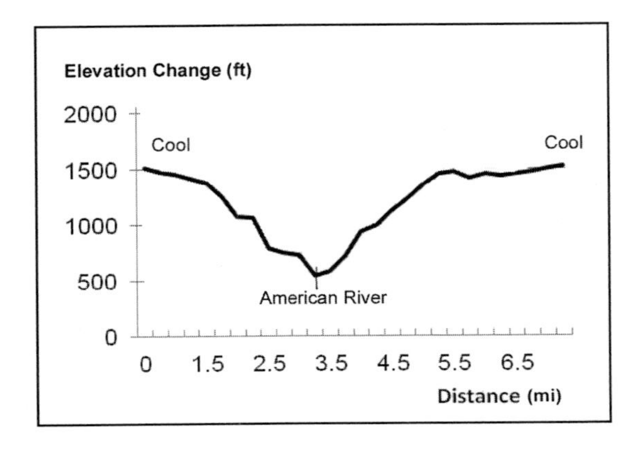

At approx 0.6 miles, the white striping on the side of the road ends. Just past this, a small trail on the right, leads to an old foundation that is visible from the road. Watch for the coyote that frequently hunts in the open areas on either side of the road. The next trail on the right is a cut off to the Olmstead Loop Trail. As you come to the next open area on both sides of the road, the road bends to the right and starts to rise.

Just before the road starts to bend to the right and climb, (N38-52-576; W121-02-076) turn right on the wide trail heading slightly downhill. Almost immediately, you come to a junction with the Olmstead Loop trail (see separate trail guide) going left and right, the Auburn Cool Trail (a return option) goes up the hill at a 45 degree angle, and the unmarked Salt Creek Trail is straight ahead. This is the official start of the Salt Creek Trail. It is the only trail going downhill from this junction (N38-52-596; W121-02-132). Follow it as it descends towards the Salt Creek Canyon through a typical shady foothill woodland.

In about a half mile, another trail coming from the Olmstead Loop, closer to Salt Creek, enters on the right. The trail downhill now opens out as you reach an area cleared during construction. The rock debris piled on the side of the trail provides an opportunity to examine the different types of rocks dislodged from the canyon. Blast holes are visible in some of the boulders. Views of the homes and trails in the Robie Pt, and Maidu sections of Auburn are visible across the canyon.

Soon the PCWA pumping station and the man made rapids that replaced the sealed, and now

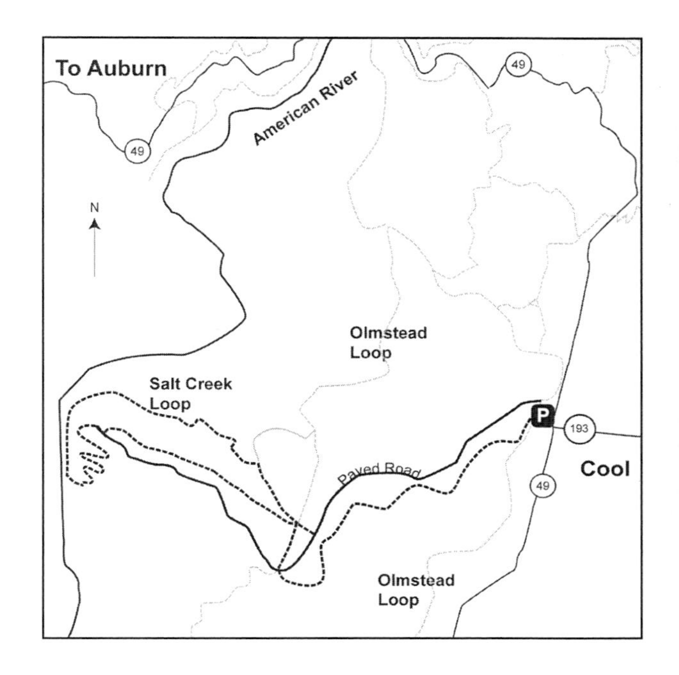

Kayakers at the artificial rapids (Rocky Island Rapids) next to the pump station.

hidden diversion tunnel will capture your attention. Look downstream on the opposite shore and note the steel netting bolted to the canyon walls to try to prevent decay. Cement was installed into many of the fissures in the rock to strengthen the canyon walls during dam construction.

The trail crosses a flat gravel area before bending to the left to start its ascent. The climb uphill is via a series of switchbacks that cut across the terraces carved into the hillside during the construction period. After about a mile, and a 600 ft elevation gain, you are at the rim and the paved road. In less than 0.4 m look for the Auburn to Cool sign and wide trail on your left.

Return Options:
The mileage difference between these three options is minimal.

a) Continue on the paved road back to the parking area.

b) Follow the wide Auburn to Cool Trail back to the initial junction with the Salt Creek Trail. At that junction – go right on the Olmstead Loop Trail for 0.2 m, cross the road and at the T, go left onto the unsigned Knickerbocker Creek Trail.

c) Follow the paved road back to the Olmstead Loop (about a mile) and take the Knickerbocker Creek Trail back. Look for the Olmstead Loop trail on your right. It is about 0.2m before the point where you initially left the road just below the crest of the hill. Go right and in a few steps at the T, go left.

The Knickerbocker Creek trail parallels and comes close to the paved road three times. It provides a relaxing trail experience as it follows the creek's bends and turns. Whenever you come to a trail junction, stay left. The trail crosses the creek two times. The first crossing has a thick board for hikers and bikers on the left.

The last crossing, less than 0.3m from the parking area, has stacked rocks to assist hikers to cross in the rainy season. The trail to the right is usually wetter and harder to cross.

Stagecoach Trail (#36 on the ASRA Topo Trail Map)

Distance: 2 miles; 1½ hrs. up, ¾ hrs down (hiking)

Difficulty: Moderate up, easy down

Elevation Change: +/- 750 ft. (see below)

Trailhead/Parking: (N38-55-010; W121-02-207)

Trailhead is at confluence area, 1¾ miles south of ASRA Park Headquarters. Take Hwy 49 from Auburn south to Old Foresthill Road at the bottom of the canyon. Continue straight for ¼ mile and park on the left. Trailhead is just beyond parking area at green gate (#137) just to the right of a bulletin board, vault toilet and shade shelter.

Description

This historic trail offers great bird's eye views of the confluence area and American River canyon. Its gradual gradient offers a good aerobic workout climbing 800 ft in 2 miles from the confluence to the top where it intersects Russell Road at a gate (#138). The history of this trail dates back to the mid-1800s (see sidebar). The trail climbs through riparian corridors as well as some typical foothill chaparral areas. Wildflowers bloom in the spring in several areas. There is little shade, so take water and sunscreen on hot summer days.

The Stagecoach Trail begins at the confluence area by the bulletin boards and heads towards the high Foresthill Bridge (see sidebar). It is marked by a "Stagecoach Trail to Russell Road" sign. The first ¼ mile is the steepest, but it provides good views of the North Fork American River, at several spots on the right.

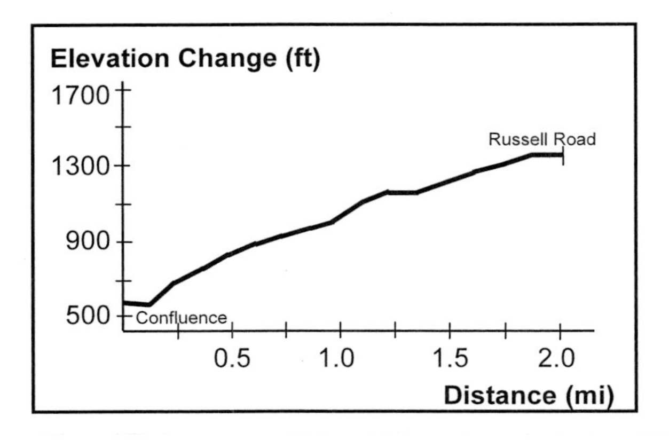

At about ¼ mile, the trail turns sharply left at the "Stagecoach Trail" sign. As you pause to catch your breath, you can almost hear the clatter of the stagecoaches that once traveled this road.

A short distance up the trail, the graceful arches of the Mt. Quarries RR Bridge come into view on the left. A little further along, the appearance of ponderosa pine, big leaf maple, interior live oak, blue oak, willow, and Himalayan blackberry bushes signal the first of several riparian corridors on the trail. Here you can see water running year round, unlike other spots on the trail where it is only visible in winter and spring.

A little further on, the steep, narrow, hiker only Tinker's Cut-off intersects on the left (see separate trail guide). A sign marks Flood Trail on the right. At the 1-mile point, a bench provides an opportunity to enjoy a great view of the confluence area and the Mt. Quarries RR Bridge. Turkey vultures are often seen here, flying low over-head, riding the thermals.

A few paces beyond the bench on the left is the narrower and unsigned Manzanita Trail (for hikers and bikers). It eventually makes its way to a junction that leads either to Canyon Drive or to the ASRA Park Headquarters.

Continuing up Stagecoach Trail, you pass open areas where California poppies cover the slopes in early spring. You also pass more riparian corridors, one of which has a small waterfall in winter and spring. Just before arriving at the end of the trail, you pass Upper Stagecoach Trail as it intersects on the right. At the top of the trail, pause to catch your breath before returning to the confluence back down Stagecoach Trail or via an alternate route (see below).

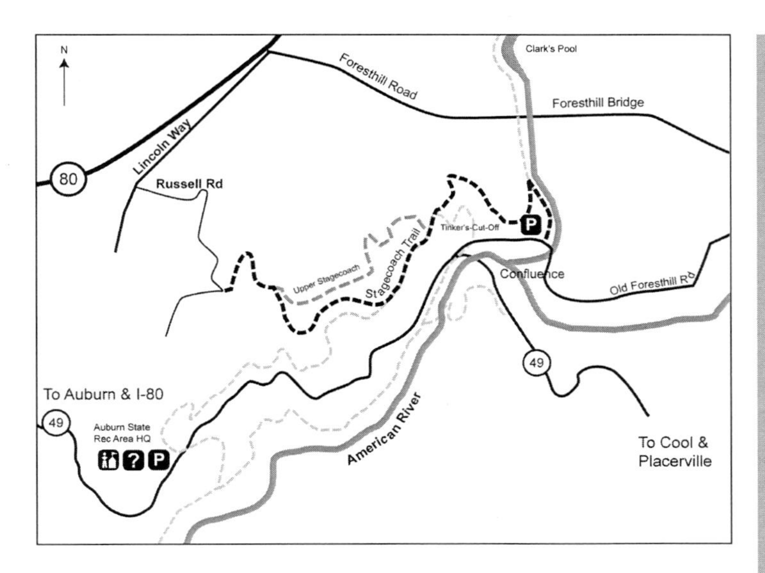

Did You Know? – Stagecoach Trail was originally a toll road built in 1852 known as Yankee Jim's Turnpike and later as Old Stage-coach Road. The original road led to a toll bridge over the North Fork American River just upriver from Clarks Hole, and from there, to the towns of Yankee Jims and Iowa Hill. In the late1800s, Yankee Jims was a popular mining area, and Foresthill was yet to be developed. In 1875, the original toll bridge was replaced with a wooden covered bridge In the 1870's, tolls on the bridge ranged from 6¢ for a cow to 50¢ for a horseman and $1 for a wagon and two horses. There were numerous stagecoach hold-ups along this trail, most notably in 1873 and again in 1880.

Did You Know? – The 2,248-ft long Foresthill Bridge was de-signed to span the reservoir that would have resulted had the Au-burn Dam been completed. (Work on the dam was discontinued in 1976.) Water was expected to reach the top of the cement piers, but today the bridge towers 730 feet above the river, making it the tallest bridge in California. It was opened in 1973 with much fanfare and has been featured in numerous movies and commercials, and it has been the site for many stunts – both legal and illegal.

Alternate Return Routes (see separate trail guides)

About ¼ mile down the trail, take Upper Stagecoach Trail to the left (hikers and bikers only). This narrow trail is not as well graded as Stagecoach Trail, but it offers nice canyon views from two wooden chairs at the top of the hill. A short distance beyond a bench, facing the Foresthill Bridge, is a fork and sign noting Flood Trail straight ahead and Stagecoach Trail to the right. Turn right and descend through a grove of manzanita, or "little apples" – tart but edible fruit popular with the animals that live in the area. On the left, remnants of a stone foundation, along with non-native plants such as lilac, a field of periwinkle and apple and fig trees provide evidence of a former home site. Continue straight past the sign for Mossy Rock Trail on the left until you come to Stagecoach Trail. Turn left to return to the confluence area.

Just ahead on the right, Tinker's Cut-off Trail offers an alternate return route for hikers. While the top and bottom are a bit steep, it is a cooler hike on hot summer days as it switchbacks through a dense, tree-lined riparian area with a small pool and year-round waterfall.

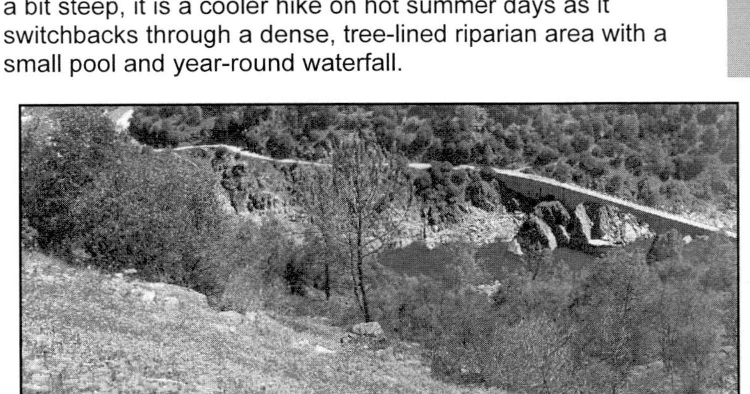

One of many beautiful canyon views from the Stagecoach Trail.

Stevens Trail (#37 on the ASRA Topo Trail Map)

Distance: 3.2 miles to river; 1½ hours down, 3 hours up (hiking)

Difficulty: Easy down, moderate up

Elevation Change: +/- 1,150 ft. (see below)

Trailhead/Parking (N39-06-376;W120-56-837)

Parking is on N. Canyon Way in Colfax. From Auburn, take I-80 east to the 2nd Colfax exit and turn left on the frontage road, N. Canyon Way. Continue 0.6 miles; turn left into a small parking area. Look for 'trail' sign marking the trailhead.

Description

This very well maintained trail is one of the most popular hiking and biking routes to the North Fork American River, in part because of its easy access from Colfax. However, this is a remote trail, and hikers are advised to carry a whistle or hike with a friend. The trail passes through oak, laurel, douglas fir, and huge stands of manzanita. The wildflower displays here are rarely matched elsewhere in ASRA, particularly during April and May. Take plenty of water and use caution, as the trail is hot during summer months and steep at times with narrow passages. Poison oak can be found along much of the trail.

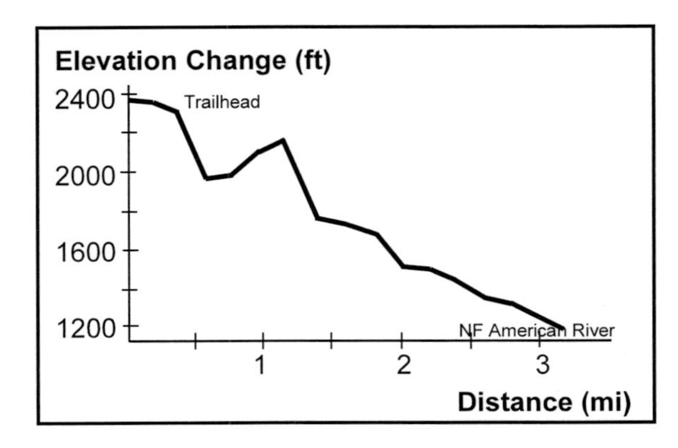

The Stevens Trail starts at the parking area on N. Canyon Way just north of Colfax. After a short distance, it intersects an unmarked firebreak road; turn right on this dirt road. At the ¾ mile point, the trail intersects another firebreak road; bear to the left and look for a metal 'trail' sign a few steps to the left. Soon thereafter, the trail forks again and you have a choice of going either right or left. A trail sign here directs bikers to use the right hand fork while hikers go left. The two trails rejoin shortly after traversing a rocky outcropping and a creek.

A little further along the trail is a beautiful cascading waterfall, and shortly thereafter is perhaps the most distinctive landmark on the trail – an immense bed of shale cascading down the mountainside. You can also see the remains of a mineshaft, abandoned long ago. In fact, there is evidence of widespread mining activity all along this trail. Above the bed of shale is the Cape Horn section of the 1st Transcontinental Railroad. Here the railroad bed was originally cut into the steep canyon wall, by Chinese laborers suspended on ropes down the steep sides. Trains still run on this section of the original 1860's roadbed.

After about a mile on the dirt road, look for a trail sign pointing to the left. From this point on, the trail enters the steep North Fork American River canyon and traverses the canyon upriver. This part of the trail offers magnificent views of the river. The Iowa Hill Bridge is visible downstream. Hand stacked rock retaining walls can be seen on this stretch of the trail, an indication of the large amount of work required to build this trail (see sidebar).

The trail meets the river at Secret Ravine, where more signs of mining are evident. The foundation of an old suspension bridge can also be seen, complete with rusting cables. Although the trail is fairly easy down to the river, take the opportunity to rest and cool your feet in the cold river before starting the more difficult trip back to the trailhead.

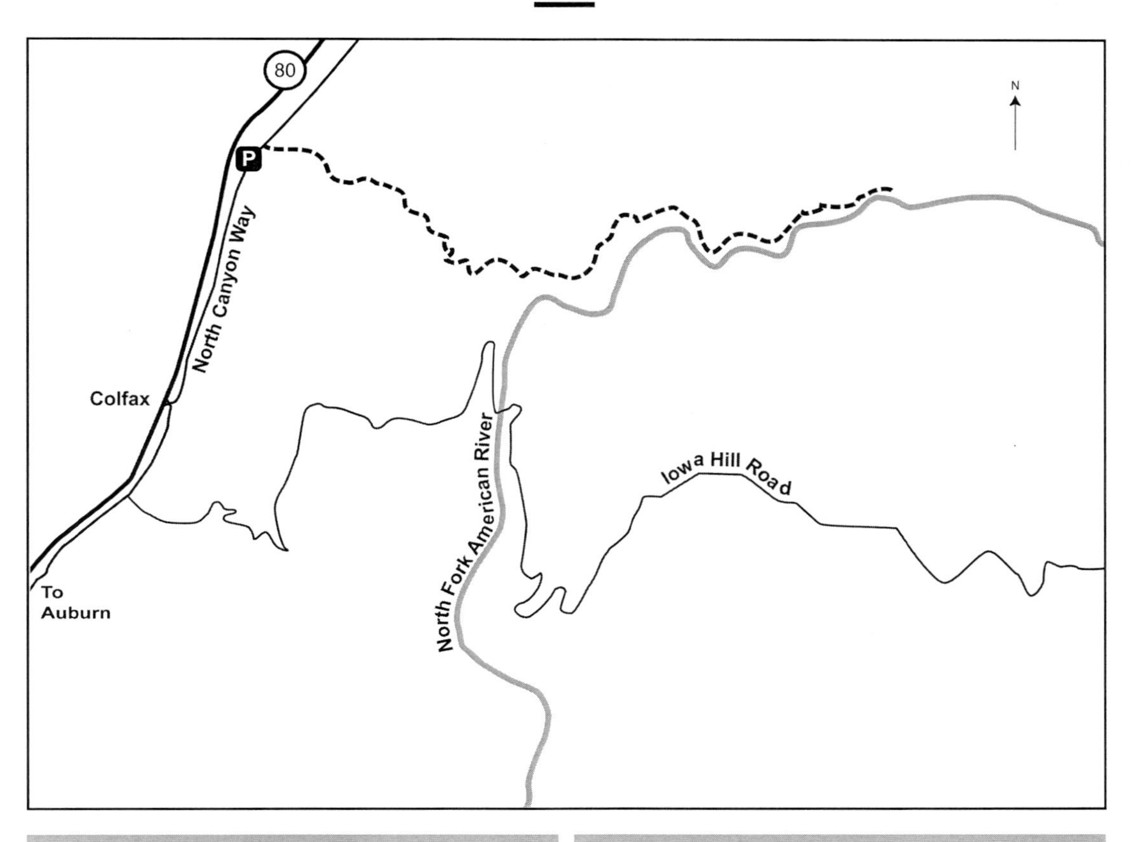

Did You Know? – The Stevens Trail was originally built by Truman Allen Stevens shortly after he arrived in California in 1859. Stevens operated a ranch in Iowa Hill and a livery stable in Colfax. He built the trail and a self-propelled cable car to transport miners and animals across the North Fork, for which he charged a toll. Although the trail can be found on a 1900-era USGS topographic map, it wasn't popular among locals until Boy Scout Eric Kiel charted it in 1969. It is now listed on the National Register of Historic Places.

Left: A view of the North Fork and the Iowa Hill Bridge from Stevens Trail.

Right: The historic Stevens Trail went to Iowa Hill and crossed the North Fork on this cable bridge, now gone.

Tamaroo Bar Trail

Distance: 1.7 to 2.3 miles one way (depending on route selected)
45 min down, 1 hr up (hiking);

Difficulty: Easy to Moderate down
Moderate to difficult up.

Elevation Change: +/- 750 ft. (see below)

Trailhead/Parking: (N38-53-358; W121-04-131)

Trailhead is at Overlook Park behind the Gold Country Fairgrounds. Take Hwy 49 and Lincoln Way through Auburn to Auburn-Folsom Road, go past the fairgrounds and turn left on Pacific Ave. Turn right after 0.5 mi into the large Dam Overlook parking area. After parking, chose one of 3 trailheads all of which begin here.

Description

This trail descends from the overlook parking area to the American River past many artifacts highlighting the construction done in the 1970's for the proposed Auburn Dam. It passes through a recovering but natural looking landscape. It offers views of the terracing on both sides of the river from the construction work as well as a view down on the man-made rapids around the permanent pump station that replaced the diversion tunnel which was closed in 2008 when the river was returned to its natural channel. The trail provides views of the American River up and down stream, as well as access to the river itself.

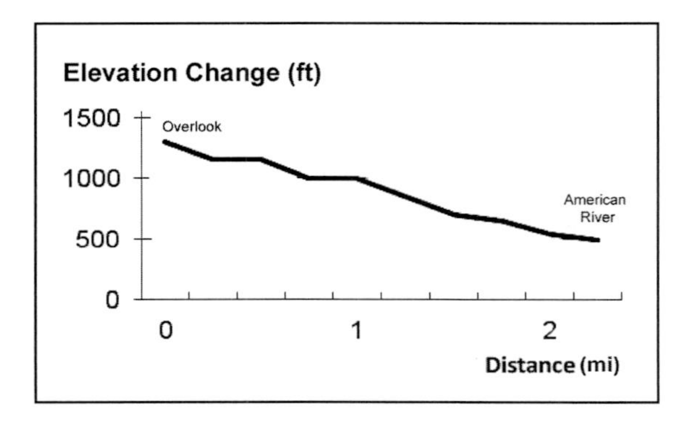

Tamaroo Bar Trail offers a chance to see the changes wrought on the land by both man and nature. It offers a variety of experiences and can easily be made into a loop since there are 3 trailheads, described below, that eventually join and lead to Tamaroo Bar and rapids.

a) From the parking area, return to the entrance road and go left for a few feet. Go through Camp Flint Gate #141, and follow the old paving down to the paved access road for utility vehicles that goes down, past the transformers, to the pump station. The road descends steeply towards the river via a series of switchbacks. About 1 mile down the paved road, look for the fairly wide trail on the left. It is where two power poles are located, one on each side of the road. In 0.1 mile the shorter more direct route from the parking area enters on the left, marked by a brown pole and trail sign. (N38-53-226; W121-03-746) This route is longer and more open but has better footing.

b) To reach this point by a more direct route, (0.5 mile) walk to the end of the parking area by the flagpole. Just to the right a trail goes down to the chain link fence. Follow the fence to the left to the opening and head downhill. Soon the trail enters a narrow gorge like area with a small creek which soon cuts across the trail. At approx 0.4 m at the unmarked T junction, go right. Soon you will see the wide main Tamaroo Bar Trail below you. When you reach the wide former road, go left. (N38-53-226; W121-03-746) This shorter route has a short steep section with loose rock that may be tricky, when wet and muddy. It is more shaded than the road and the shortest route.

Both these routes offer views of the terraced roadways up and down the canyon that provided access for construction equipment, as well as the newly built permanent pump station and manmade rapids which replaced the now buried diversion tunnel. Continue on the path for approx 0.4m to where the route down from the Robie Pt Fuel Break / Western States Trail (WST) enters on the left. There is a post on the right. (N38- 53-251; W121-03-322)

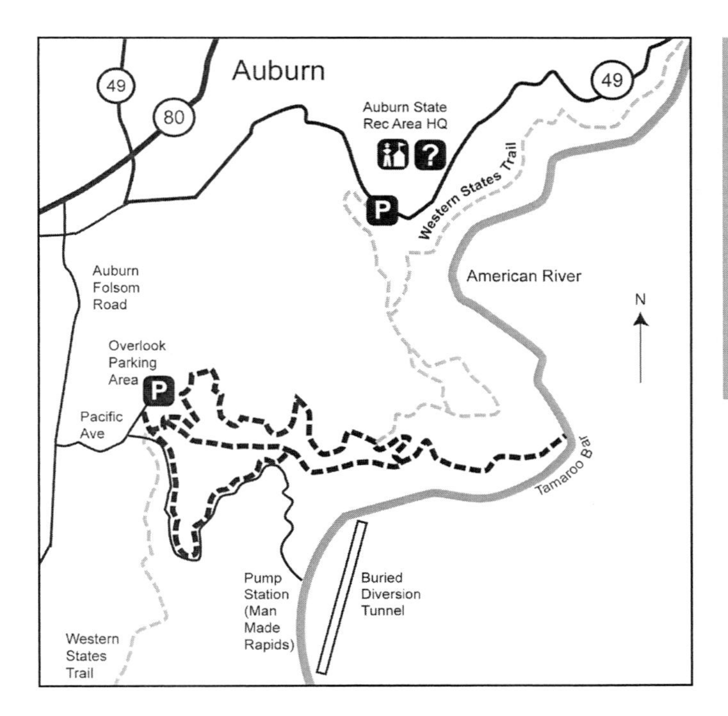

Did You Know? –Tamaroo Bar is named in honor of the ship Tahamaroo. The ship carried the first miners (aka -49'ers), via Cape Horn, to San Francisco in 1849. Some of the passengers on the ship became miners at this location. By 1852 the first "h" and "a" in the name had been dropped by the local newspapers and the generally acceptable spelling became Tamaroo.

Former buoy line for coffer dam.

c) To reach this point via the Robie Pt Fuel Break/ WST, (see separate guide) from the parking lot return to the road and go right into the equestrian staging area below the rest rooms. Go through the fence by the bulletin board, turn left and go straight past the Pioneer Express sign. This shaded section of the WST trail goes gently downhill, and across a creek. Ignore the trails that go uphill on the left and stay on the Robbie Pt Fuel Break / WST trail. In approx 1.1 miles there is a U shaped trail junction with the Robbie Pt / WST (N38- 53-312; W121- 03-374) going up to the left. Take the downhill trail on the right. This route will switchback its way to the lower Tamaroo Bar Trail in approx 0.3m. At the T junction, marked by a sign post, go left onto the Tamaroo Bar Trail.

Now that all three access routes have merged, the trail crosses a terraced wide open fill area before it narrows as it begins to drop towards Tamaroo Bar and rapids. When the dam was being built, this area was behind the coffer dam built to hold water out of the construction zone.

Soon you'll note a series of posts on the left with white plates and black markings, which resemble piano keys. These were the depth markers for the water held behind the coffer dam. Note also the overhead wire with rusting metal drums. This is the former buoy line beyond which boats were not allowed. Shortly the river and Tamaroo Bar, a once highly profitable gold site, will be in view.

Formerly the gravel bar was connected to the bank but the river created a deep wide channel separating it from the shore. Just before reaching the river, there is a small trail going off to the right. It parallels the river downstream towards the pump station. Continue to the left towards the boulder strewn Tamaroo Bar rapids. Be careful as the trail is close to the edge of a steep sand bank which is being eroded by the river. Depending on the water level, it is fun to sit on the boulders alongside the rapids and enjoy the river while deciding which return route to take.

Tinker's Cut Off Trail (#38 on the ASRA Topo Map)

Distance: 0.3 miles one-way; ¼ hours (hiking)

Difficulty: Moderate to Difficult

Both top and bottom sections have been severely rutted by unauthorized bicycle use.

Elevation Change: +/- 350 ft. (see below)

Trailhead/Parking (N38-54-969; W 121-02-431)

The trailhead is 1-¾ miles south of ASRA Park Headquarters in the confluence area. Take Hwy 49 from Auburn south to Old Foresthill Road at the bottom of the canyon. Park as close to the Hwy 49 Bridge to Cool as allowed.

This unmarked trailhead is almost directly opposite the Hwy 49 Bridge on the Placer County or north side of the river. Looking uphill a clear, sharp cut in the bank is visible.

The trail may also be accessed off of Stagecoach Trail. The upper trailhead for Tinker's Cut Off is just a little less than 1 mile from the confluence via Stagecoach Trail. (N38-55-049; W121-02-557)

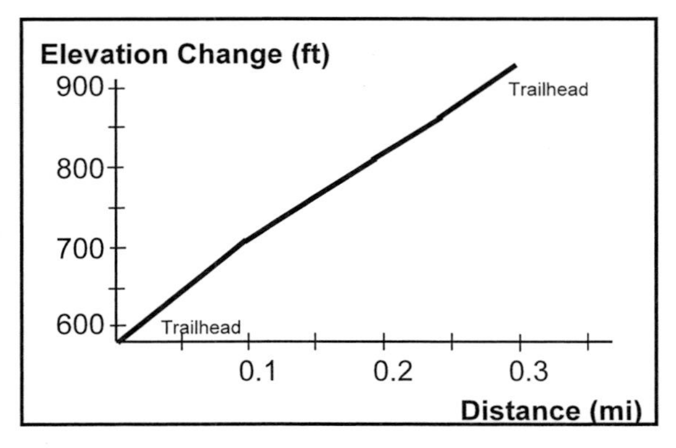

Description

This single file trail offers an alternate route to and from Stagecoach Trail from the confluence area. It follows a creek drainage through a tree-lined riparian area with a small pool and waterfall that lasts year round. It is a popular cut off for hikers in the summer months, providing welcome shade that is not always found on Stagecoach Trail.

Tinker's Cut Off may be hiked up and back, but most people use it to either go up to Stagecoach Trail from the confluence or as a cut off from Stagecoach back to the confluence area. The bank at the bottom has been badly torn up by unauthorized biker use. Going straight up the hill, the trail crosses a gully via a wide thick wooden plank.

As you continue up though the tall trees and dappled shade, moving water may be heard and seen bubbling over the rocks and boulders in the creek bed on the right. Just before the trail turns sharply left and away from the creek, not far below Stagecoach Trail, behind the steep bank, there is a small opening on the right leading to a pleasant mini-waterfall in the creek. This riparian oasis can be a cool spot for a break on a hot day.

As you leave the creek, the trail begins a series of switchbacks leading to Stagecoach Trail.

Upon reaching Stagecoach Trail, (see separate trail guide) the choices are to return back to the start; turn right and return to the confluence area via the slightly longer but less steep Stagecoach Trail, or go left and continue up Stagecoach Trail to Russell Road before returning to the confluence area.

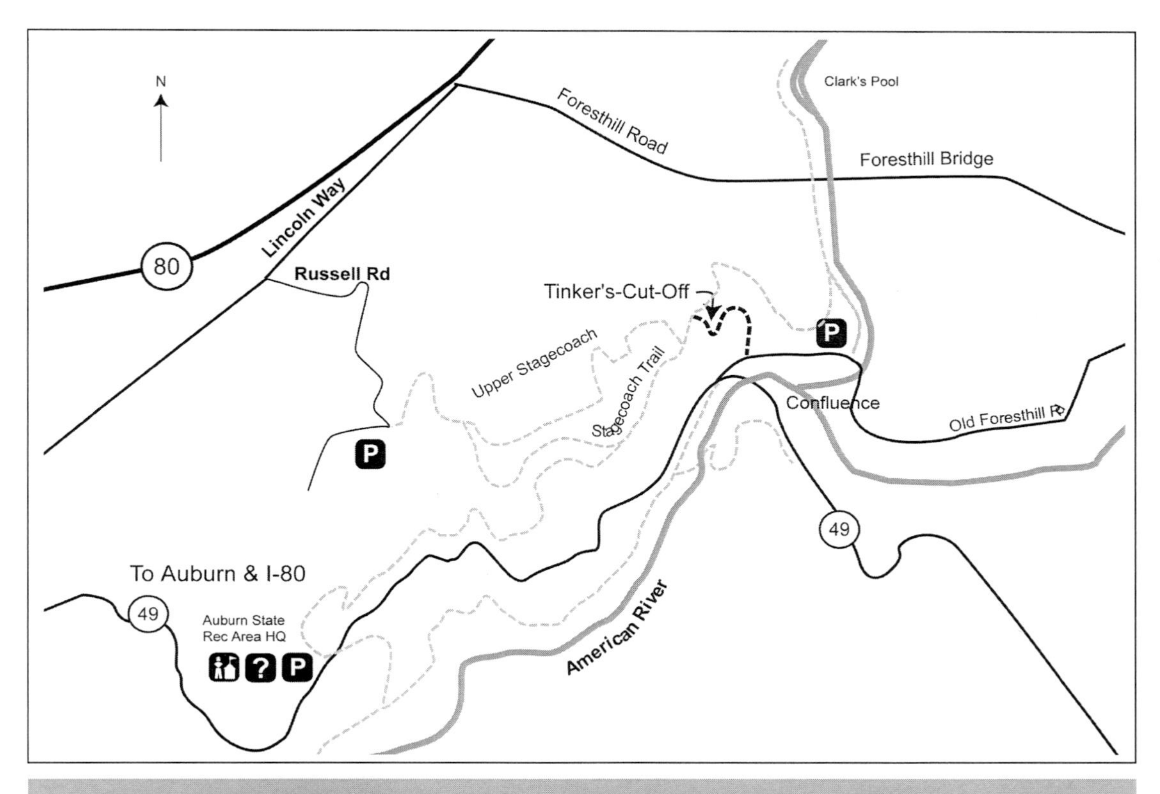

Did You Know? – In recent times, new trails in ASRA must be environmentally approved and built to state park trail standards. In the past, many trails were built and maintained by individuals.

One local couple, Bob and Harriet Stephens, using their own tools, cleared and created several of the popular routes in use today. This is one of the trails they created. Tinker's Cutoff is named for Harriet, (Bob's nickname for his wife).

Left: view of the confluence area from Tinker's Cut Off.

Right: pool along the trail.

Upper Stagecoach Trail (#39 in the ASRA Topo Trail Map)

Distance: 1 mile one-way; ½ hr (hiking)

This does not include the distance walked on Stagecoach Trail to the trailhead. Access to this trail is from Stagecoach Trail.

Difficulty: Easy to Moderate

Elevation Change: +/- 300 ft. (see below)

Trailhead/Parking (N 38-54-683; W121-03-257)

Parking is off Russell Rd. From I-80 exit at Foresthill. Turn right (west) on Lincoln Way. Follow Lincoln way for about ½ mile and turn left onto Russell Road. Drive slowly as many residents, children, and pets may be walking or biking along this single lane road. After about ½ mile, there is a small parking area on the left as you round a curve to the right. Trailhead is behind the parking area beyond green gate (#138).

Description

Trail goes from Stagecoach Trail to Stagecoach Trail. While not as wide, level or well graded as Stagecoach Trail, it provides another perspective on the canyon, and has some shade. On clear days it is possible to see the snow-capped Sierra Nevada as well as views both up and down the American River Canyon. There are two locations where there are benches from which to enjoy the views along the way.

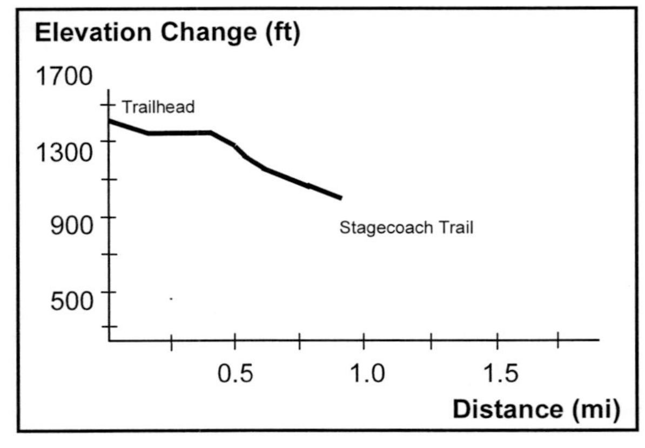

The trail crosses several small oak meadows. It also provides a good view of the imposing Foresthill Bridge (see sidebar) and some opportunities to watch hikers and bikers below on Stagecoach Trail.

Upper Stagecoach Trail is usually used as an alternative route to or from Stagecoach Trail (See Sidebar) or as a shorter hike from Russell Road. It can be hot in the summer.

From the top of Stagecoach Trail (off Russell Road) proceed down Stagecoach Trail for just over a ¼ mile. Upper Stagecoach Trail begins at the sign on your left. (N 38-54-730; W 121-03-142). Go left up the hill at the sign for Upper Stagecoach.

There are several unofficial unmarked trails going up to private property and homes near the canyon rim as well as others going down that are used mainly by bikers to access Stagecoach trail below. The main trail is easy to follow.

Shortly after reaching the top of the rise, the trail comes out of the trees and shrubs into an area of small open meadows. In less than 0.2 miles there is a nice wide wooden chair offering views of the canyon and the hikers below on Stagecoach Trail. The view encompasses Route 49 curving towards Cool and the Quarry Trail along the south bank of the Middle Fork American River as well as views up the Middle Fork. The wide steep trail rising to the ridgeline across the American River, at the end of the cable stretching across the canyon, is the Pointed Rocks Trail to Cool (see separate guide).

After passing a small meadow, but before re-entering the trees and scrubs, there is another bench facing the Foresthill Bridge.

Shortly after reentering the trees and shrubs, at ½ mile, Upper Stagecoach Trail meets Flood

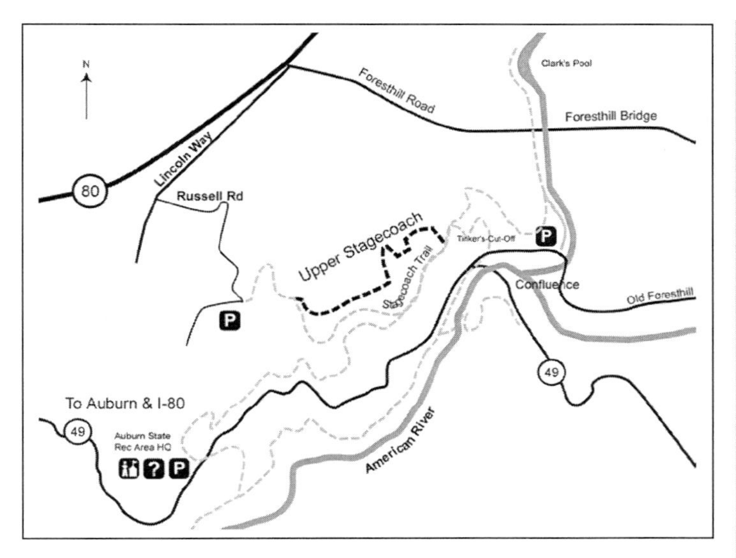

Trail. At the fork, a sign points to Stagecoach Trail to the right and Flood Trail straight ahead. This connector down to Stagecoach is ½ mile, and is usually considered part of Upper Stagecoach Trail. Do not continue straight on Flood Trail. This narrow, overgrown, and single file trail leads up to Flood Road, a residential neighborhood with no parking. The trail is largely overgrown and offers little to see.

At this point, either retrace your steps to the start or go right to return to Stagecoach Trail closer to the confluence. Those turning right will soon be following a rutted single file red clay trail through a small manzanita grove. After this, watch for a large field of periwinkle on the left. Looking behind the periwinkle the remnants of the stone foundation of a house may be seen. A few steps further, on your left, is the start of the Mossy Rock Trail. Continue straight-ahead and down. In about 0.1 of a mile you will reach Stagecoach Trail at a little past the halfway point. From here, turn right and return to Russell Road or turn left and follow Stagecoach Trail down to the confluence. (See separate guide for Stagecoach Trail.)

Did You Know? – Stagecoach Trail was originally a toll road built in 1852 known as Yankee Jim's Turnpike and later as Old Stagecoach Road. The original road led to a toll bridge over the North Fork American River just upriver from Clarks Hole and from there to the towns of Yankee Jim's and Iowa Hill. In the late1800s, Yankee Jim's was a popular mining area and Foresthill had not yet developed. In 1875, the original toll bridge was replaced with a wooden covered bridge (pictured in the kiosk located at the top of the trail). In the 1870's, tolls on the bridge ranged from 6¢ for a cow to 50¢ for a horseman and $1 for a wagon and two horses. There were numerous stagecoach hold-ups along this trail, most notably in 1873 and 1880.

Did You Know? – The 2,248-ft long Foresthill Bridge was designed to span the reservoir had the Auburn Dam been completed. (Work on the dam stopped in the late 1970's.) Water was expected to reach near the top of the cement piers. The bridge towers 730 feet above the river, making it the tallest bridge in California. It was opened in 1973 with much fanfare and has been featured in numerous movies and commercials and has been the site for many legal and illegal stunts.

A resting place on the trail.

Western States Trail - Hwy. 49 to Robie Pt. FB

Distance: 2 miles; 1 hour each way (hiking)

Difficulty: Easy to Moderate

Elevation Change: +/- 450 ft. (see below)

Trailhead/Parking:

East (Confluence):(N38-54-893; W121-02-388)

West (Hwy 49): (N38-54-001; W121-03-536)

East (Confluence Area by Hwy 49 Bridge):

Trailhead is on Hwy 49, 1¾ miles south of ASRA Park Headquarters. Take Hwy 49 south from Auburn towards Placerville. After crossing the American River, park on right off the highway. Walk to trailhead through the green gate (#150).

West (Hwy 49 – Robie Point trailhead):

Trailhead is accessed from parking area on Hwy 49, southbound from Auburn ½ mi (on the right) or northbound from ASRA Park Headquarters ¼ mi (on the left). **Caution: cars coming downhill are hard to see.** From the parking area behind the gate (#130), walk ½ mile of Robie Pt trail. The trailhead is a narrow trail on the left marked with a metal "Trail" or WST sign. (N 38-53-707;W121-03-426)

Description

This short section of the Western States Trail offers an easy outing with ever changing views of the American River. The trail follows the old Mt Quarries RR roadbed for most of its route. Thus it is fairly wide and level in most places. Users pass several reminders of the railroad along the way. The trail remains in sight of the river through much of its length. Several small unofficial trails provide access down to the river and its many gravel bars that serve as swimming spots.

Elevation Change (ft)

1100
900
700
500
Trailhead (Robie Point)
Park Access Trail
Trailhead (Rte 49)
Waterfall
1
2
Distance (mi)

The Hwy. 49 to Robie Pt. FB section of the Western States Trail is easily accessible and provides a scenic outing. Leaving trailhead east at the confluence, one quickly arrives at the Mt Quarries RR Bridge, which offers eye-catching views up and down river in all seasons. Nesting swallows always put on a show in spring. Be sure to stop at the far end and read the plaque, outlining the history of this graceful structure. The route continues just above the American River on river right. Along the way, the small trails on the left lead down to gravel bars that serve as swimming areas in the summer. The wider main trail continues downstream. After 0.7 miles, the first of several concrete remains of trestle abutments will be found on the left side of the trail with the date 1921 etched on the corner. Since these remnants always mark where the railroad crossed a ravine, look for its mate on the other side. At these spots, since the trestles no longer exist, the trail leaves the original roadbed and become narrower and not as well graded as it goes up, down and around the intervening ravine. Note the various dates etched in the concrete trestle abutments along the way.

After 1 mile, the trail has a steep decent and ascent to and from a pretty area popularly known as "the Black Hole of Calcutta". As the name implies, the area is always shaded and darker than the surrounding area. Mist or spray from the graceful waterfall ensures a consistently moist and cool environment. Horses splash across the boulder strew stream crossing while hikers hop from rock to rock.

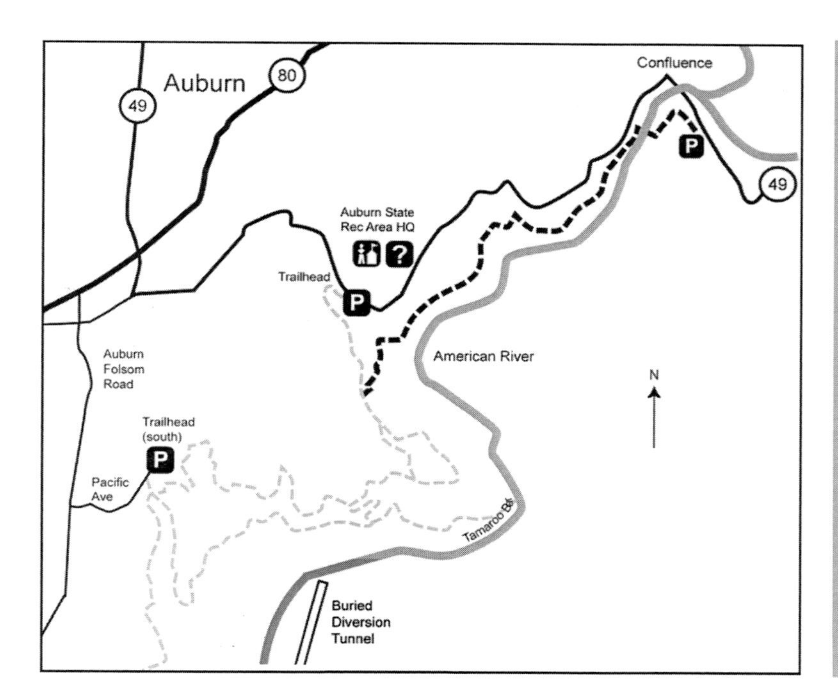

Did You Know? – The Western States Trail originally stretched from Sacramento to Utah. The Sierra Crest portion of the trail, blazed by Paiute and Washoe Indians and later used by miners, is now the route of two world-famous endurance races: the Western States 100-mile Endurance Run and the Tevis Cup Ride for equestrians. This part of the WST (from the Mountain Quarries RR Bridge to the Auburn Staging Area) is the final leg of these races. The WST mileage markers along the way indicate the distance to the finish line in Auburn.

Leaving this oasis, the trail climbs back to the railroad bed. At 1.28 miles, the Park Access Trail comes in on an angle on the right. Along the next stretch there are some nice views of the river and wildflowers in spring. At 1.5 miles, the trail passes a rock outcropping on the right and the tall somewhat pyramid shaped spire known as Eagle Rock. Don't expect eagles, but there is a good chance of finding turkey vultures soaring.

Past the rock outcrop, in the tiny clearing on the right, shaded by a large oak, are some rotting timbers. This is all that remains of the water tower that served here as a watering stop for the RR. A few paces further, at 1.6 miles, is a trestle abutment from 1917. This is as far as the trail follows the old roadbed. If you are doing this as an out and back, this is a good turn around spot. From here the trail ascends via a series of zigzags through the scrubs and trees up the hillside to the Robie Point FB Trail (see separate trail guide). This section may be muddy and slippery after heavy rains.

Reaching the Robie Point FBT, riders go left following the WST to the staging area in Auburn. Hikers often go right for ½ mile to the parking area on Hwy 49, (trailhead west) a place to leave a second car to allow a shuttle.

Rider near Eagle Rock.

Windy Point Trail (#49 in the ASRA Topo Trail Map)

Distance: 1.5 miles one-way; ¾ hours
Difficulty: Moderate to difficult. Trail is narrow, with loose rock and steep drop-offs
Elevation Change: +/- 550 ft. (see below)

Trailhead/Parking (N39-05-420; W120-55-259)

Trailhead is on Iowa Hill Road. From Auburn, take I-80 east to Colfax. Exit at the Colfax, Grass Valley, Hwy 174 exit. Go right at the exit; at the stop sign, turn right onto Canyon Way. In approx. 0.3 miles turn left on Iowa Hill Road. Drive approx 3.1 miles down Iowa Hill Road and across the North Fork American River. After crossing the bridge continue for approx 0.8 mile through the small camping area and up the road towards Iowa Hill. Several large boulders and a very small sign that simply says "trailhead" mark the trail. There is limited parking on the right shoulder just before and at the trailhead as the road bends to the left and away from the river. **Caution: Iowa Hill is a narrow paved road with no guardrail.** Drive slowly and watch for approaching vehicles since some places are barely wide enough for two vehicles to pass.

Description

This short but challenging trail affords some spectacular views of the North Fork Canyon and when timing is right a spectacular wildflower display. The trail traverses sections where the whole hillside is carpeted with blue, yellow, purple, and gold wildflowers highlighted against a blue sky and green hills dominated by oaks and gray pine. In spring, it is a photographer's and wildflower lover's delight. Caution: watch for poison oak, as it thrives along the trail.

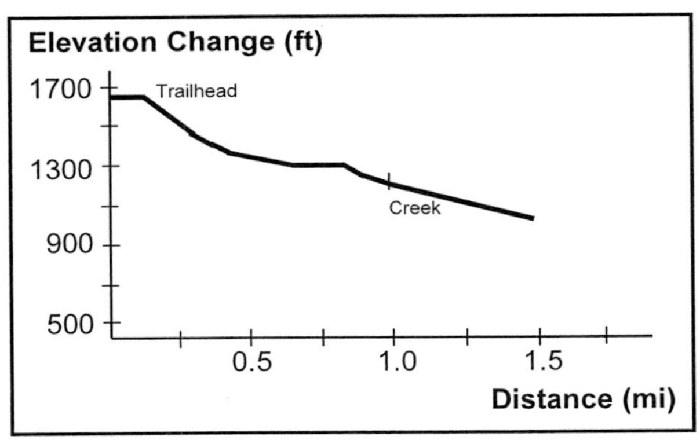

Windy Point Trail provides access to some beautiful and steep river canyon scenery. This old miners trail is a narrow single file path cut into a steep slope above the North Fork of the American River. It has some sharp drop offs and lots of loose shale and rock, which make it a challenge. This is a place to stop often to absorb the magazine centerfold vistas. Plan on spending as much time enjoying the scenery as hiking. Near the start, it is possible to look back and north across the canyon to the famous "Cape Horn" section of the 1st Transcontinental Railroad, which is still in use. Its completion was one of the extraordinary engineering and construction feats of the 1860's and the building of the 1st Transcontinental Railroad.

The trail starts out fairly wide and level in the shade of a foothill's woodland. It soon narrows and starts treading its way across a steep slope above the river gorge. The river below will be heard well before it is seen. Shortly after coming out into the open, look across the canyon and note the long thin waterfall visible just south of the Iowa Hill Road. While not visible from the road, it runs down the canyon wall almost from the rim. After an old burn area, windy point will dominate the slope above on the left. Remember to watch your footing while noting the colorful wildflower displays in the many small clearings.

Note the various shades of green on the steep canyon walls on the opposite side of the canyon. Kayaks may be visible in the river below as they exit Chamberlain Falls and prepare to enter Tongue and Groove on this class III – IV section of the river.

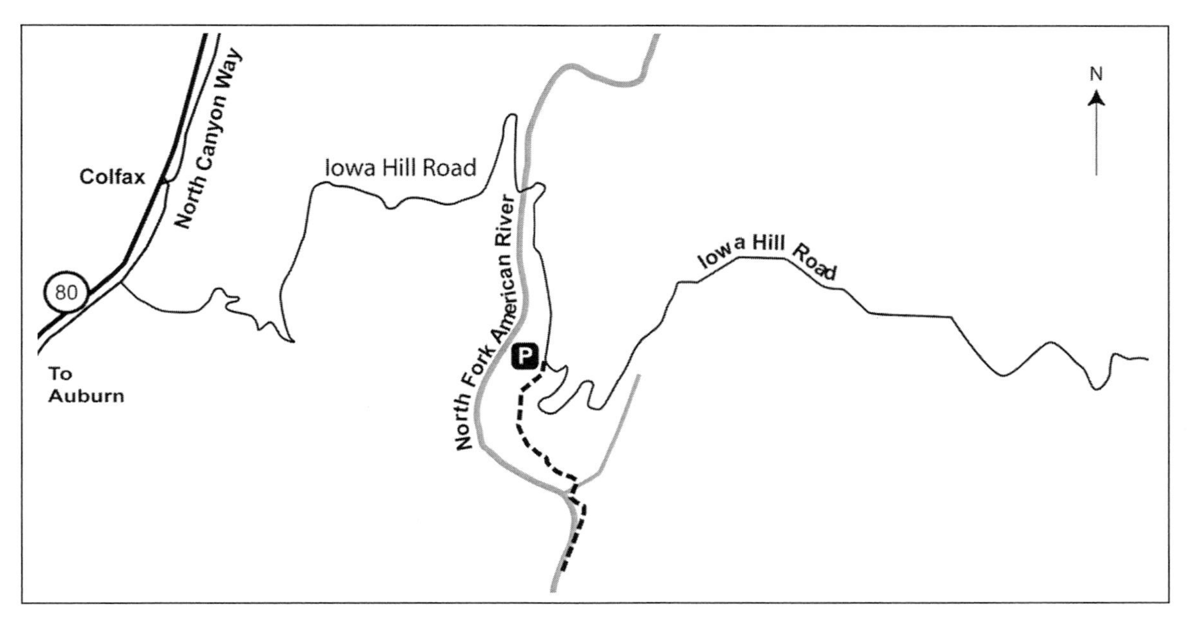

Butter and eggs, popcorn flowers and brodeas usually dominate the first section of open meadow in early spring. Another section above the trail resembles a series of rock terraces filled mainly with lupines and poppies.

In just over ½ mile, a large open slope falls away to the river. In spring this slope is often carpeted in golden yellow with blue and white accents set off by a few oaks and pines until it reaches the rapidly flowing river below.

The less frequently seen harlequin (purple and yellow) lupine can usually be found here. Soon the trail makes a short, steep descent to a wide creek. Watch out for the poison oak that lines the bank.

Depending on the water level, this may be the end of the trail. If the creek can be safely crossed, the designated trail continues on the other side for about ½ mile. There is a path down to the river at this point.

Windy Point Trail and a view of the North Fork American River.

Confluence Interpretive Trail

A Self-guided Historic Bridges and Nature Walk

Distance: 2.6 miles total
Two independent sections (0.8 mile loop and 1.8 miles out and back)

Difficulty: Easy

Elevation Change: +/- 100 ft. (see below)

Trailhead / Parking: (N38-55-010: W121-02-207)

The trail markers are small, discreet metal posts that are only a few feet high.

This trail begins at the confluence area, about 3 miles below Auburn. From Lincoln Way in Auburn take Hwy 49 (El Dorado Street) to the bottom of the canyon. Continue straight past the Hwy 49 Bridge for ¼ mile and park either on the right or left side of the road. Start the self-guided trail by walking on the south (left) side of the Old Foresthill Rd. towards the Hwy 49 Bridge marker post #1 (about 30 yards before the bridge).

Note: The information panel located on the NW side of the Old Foresthill Bridge includes pictures of some of the bridges described in this guide.

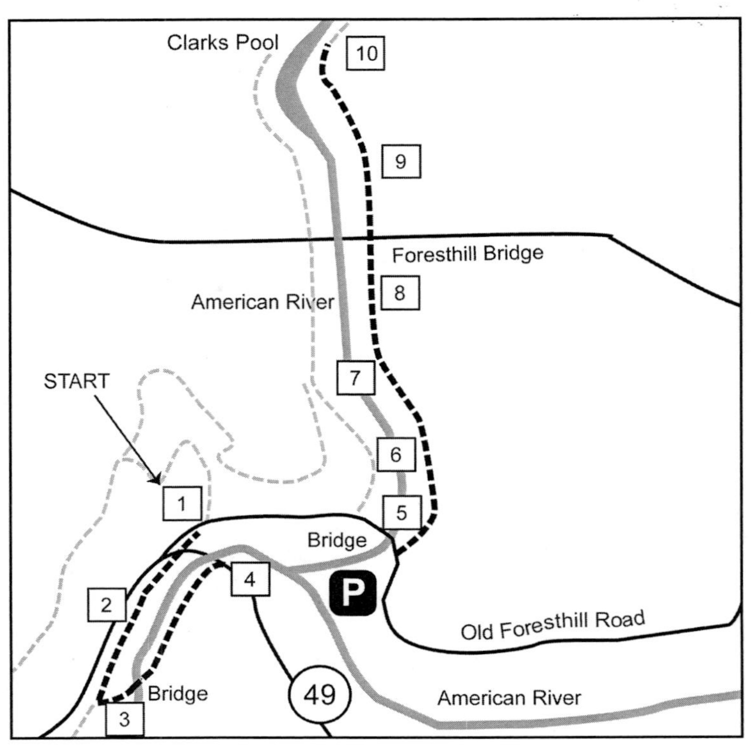

Trail description: The confluence of the North and Middle Forks American River affords a marvelous opportunity to view the natural wonders of this river canyon environment and to glimpse the past via the remains of its historic bridges. The first section of the trail goes from marker posts #1 to #4. This short 0.8 mile loop encompasses the sites of five bridges (two still remain). For the second section, which goes from marker posts #5 to #10, walk from the SE side of the old Foresthill Bridge up the North Fork to the site of Clarks Pool (0.9 mile) and back.

History at the confluence: The area near the confluence of the North Fork and Middle Forks American River below Auburn has been the site of a dozen

An aerial view of the confluence area.

bridges from the 1800's to the present. Just four remain. Native Americans (predominantly Maidu) encamped in this vicinity for hundreds of years, but were displaced by miners who scoured the canyons during the mid 1800's in search of gold. Bedrock mortars (Indian grinding holes) can be found in the rocks where the two rivers meet at the confluence. From 1848 into the 1850's, the American River from Sacramento to the high Sierra Nevada Mountains was lined with mining camps and miners. Stage roads and bridges were constructed throughout the area to facilitate travel between Sacramento, Auburn, Foresthill and other mining camps. Remains of many of these routes and camps are still visible today.

Gold at the confluence

None of the rocks along the banks of the rivers in the confluence area contain gold. Cooling intrusive igneous rocks like those in the high Sierra Nevada Mountains form gold. As one of the last minerals to crystallize out of a magmatic melt, gold often forms at the same time as pure quartz - thus miners look for veins of quartz to give them clues to where to find gold. Gold is also found in the sediments along the river bottom. Known as placer gold, it erodes out of other locations and is washed down stream by the action of running water. Today panning or dredging yields sparse amounts of gold. The area's real riches are in its natural beauty!

The Mountain Quarries RR started operation in March of 1912.

Geology at the confluence: Most of the rocks at the confluence were formed about 160 million years ago, during the Jurassic Period. At that time, the western coast of North American was located near the current California-Nevada border and the confluence area was part of the ocean bottom forming the continental slope. Over the millennia, fine-grained muds and silts were deposited over the basaltic ocean crust.

California has been on the boundary of the North American and Pacific Plates for about the last 200 million years. The result has been that the state has been subjected to earthquakes, volcanoes and high heat flow for a long time, and the rocks have been squeezed, heated and faulted into contorted metamorphic slates, serpentine and other formations. About 50 million years ago, the actual plate boundary ran close to the Auburn area and the Foothill Fault Zone was formed. Many of these faults can be seen in the area of the confluence and the site of the proposed Auburn Dam. The movement of the plates caused large cracks, or faults to open in the ocean crust, crushing the rocks along the edges of the faults. Although the fault system in the Auburn area is no longer active, it left zones of fractures and crushed rocks, which were more easily eroded than the surrounding rock. When the Sierra Nevada Mountains began to rise about 5 million years ago the newly forming river systems followed the fault zones. The steep sides of this river canyon, with broken rocks in the bottom and the abrupt right angle turns in the rivers, are indicators of this tendency of rivers to follow faults.

Vegetation: The vegetation that is visible from this trail is either typical of the drought-tolerant woodland on the canyon walls or the wetter riparian zone along the river's edge.

Dominant trees in the woodland areas include: interior live oak, canyon live oak, black oak, blue oak, madrone, foothill pine, douglas fir, and California bay laurel. Typical understory plants in the area include: toyon, manzanita, philadelphus, styrax, poison oak, redbud, and coyote bush.

Dominant trees in this riparian area include: alder, cottonwood, willow, valley oak, ash, big-leaf maple, and locust. Common plants in the under story of the riparian area include: blackberries, button-brush, ferns, wild grapes and sedges.

An exceptional variety of wildflowers can be found along this interpretive trail from February through July with the peak time usually in early to mid-April.

Marker Post #1
Lyon's Bridge - 1865-1930

History: Prior to the building of the first bridge at this location, early miners and pioneers forded or ferried across the river. In the 1860's, the need to get easily from one side to the other led to the building of the first bridge at this site in 1865. A small town developed around the bridge including a saloon, store and dance floor, which you can see in the photo. Known as the "Lyon's Bridge," it was built with private capital and operated as a toll bridge. One-way rates ranged from $0.25 for a footman, $1.25 for a two-animal team, $0.10 each for loose stock and $2.50 for an eight animal team. The Lyon's Bridge was in operation until 1930 -- longer than any other bridge at this site to date. There have been three more bridges spanning the river since then, including the current one. The state took over the road and bridge in 1921.

Plants: This is an area that has been disturbed by roadways and other construction and is dominated by exotic (introduced) plants including fig and locust trees, Himalayan blackberries and star thistle. About 50 feet down the trail (heading under the bridge) on both sides is poison oak. All parts of this plant -- leaves, stems, and berries – cause a painful rash for most people.

Geology: Notice the steep canyon slopes, which indicate that the American River is still cutting through the uplifted mountain range. Also note the varied colors of the rocks in this area (green, reddish, dark brown, gray), which reflect the different mineral compositions of the original mud, silt or basalt.

Behind the #1 marker post the trail descends towards the river and proceeds under the current Hwy 49 Bridge. As you walk toward stop #2, notice the remnants of the 1948 bridge down river from the current bridge at the water's edge. It was deposited there by the flood of 1964.

In 1911, the confluence was the site of a cable bridge, saloon, store and outdoor dance floor.

Marker Post #2
Mountain Quarries RR Bridge

The path you are now on was built as a connector trail when the Mt. Quarries Bridge was closed to traffic in 1997, while undergoing a million dollar repair project.

History: Construction on the Mt. Quarries Company RR Bridge (aka No Hands Bridge) was completed in 1912 at a cost of $300,000. It was one of the largest arched concrete bridges in the world at the time. It was built by the Mountain Quarries Company and the Pacific Portland Cement Company to haul limestone from their quarry on the Middle Fork American River to the Southern Pacific RR in Auburn. Several hundred men worked to complete the 170-foot long, three span reinforced structure.

Trains ran four times daily, using two engines, along the standard gauge line that operated until 1940. There were no turn-a-rounds so the engine backed all the way down -- about seven miles. In 1942, the quarry ceased operations and the tracks were taken up for scrap iron for WW II.

This bridge has withstood numerous floods including a disastrous one caused when the Hell Hole Dam upstream on the Rubicon River, then under construction, gave way on December 23, 1964. The flood washed out the then newer 1948 Hwy 49 Bridge upstream. Traffic on Hwy 49 was routed over the Mt. Quarries RR Bridge while a new Hwy 49 Bridge was built.

Limestone was hauled from the quarry to the main line in Auburn from 1912 to 1940.

Plants: Most of the plants that you see in this area are non-native annuals. The large plants and trees visible on the opposite side of the canyon are almost all natives. Below this stop you can see native plants beginning to dominate formerly disturbed areas. These include: poison oak, buckwheat, several species of oak and penstemons.

As you walk towards the next stop (#3) you will move through a near tunnel of native plants – live oak, toyon, coyote bush, buckwheat, poison oak, clematis, and California bay laurel.

Geology: Looking across the river you can see straight lines (actually fractures or natural breaks) in the rocks that occur in different orientations indicating that they have been created at different times during the past 160 million years.

Marker Post #3
A Local Nickname

Many wonder how the Mountain Quarries RR Bridge received the nickname "No Hands Bridge." During the early years of the Tevis Cup Trail Ride from Squaw Valley to Auburn, people often commented that it must be dangerous to ride across the bridge without any guardrails. In jest of these fears, veteran rider Ina Robinson would drop the reins on the neck of her Arabian horse, hold out her arms, and say "Look, no hands". Gradually the expression became a common name for this bridge.

The nickname "No Hands Bridge" was started by horseback riders.

Note the construction photo. The large ramp in the center of the bridge was for pouring concrete. The quarry plant, railroad and bridge were constructed simultaneously at a total cost of about 1 million dollars (about 16 million dollars in 2005).

Construction of the Mountain Quarries RR Bridge was started in 1911 and completed in March of 1912.

The bridge was listed on the Register of National Historic Places in 2004. Please see the plaque and history display at this stop on the west end of the bridge.

Plants: Look for the undisturbed areas of heavy tree cover on the opposite side of the canyon. How many different species can you see? A partial list includes foothill pine, ponderosa pine, douglas fir, interior live oak, black oak, valley oak, blue oak, buckeye, cottonwood, willow, bay laurel, madrone, alder, ash, locust, big leaf maple and fig.

Proceed across the bridge and walk to Hwy 49. Watch closely for traffic as you cross to the other side of the highway and step across the safety rail. The marker for stop #4 is to the left near the bank.

This is the first bridge at the confluence built in 1865. Today's Hwy. 49 bridge sits right next to this site. The original rock abutment of this bridge can still be seen on the bank.

Marker Post #4
Lyon's Bridge - 1865

History: The photo above depicts the first bridge at this location in 1865 (before it was refurbished in 1917 with rooftops over the cable towers). It lasted 65 years -- longer than any other at this site. This bridge was a privately built and tolls were charged for its use.

Notice the small village in this photo. Can you see the prominent rock below the abutment? Do you see the same big rock on the bank across the river today?

The state took over the bridge in 1921. In 1930, the original bridge was dismantled and replaced by a suspension bridge. In 1948, a modern two-lane bridge replaced the suspension bridge, which was becoming unsafe. It provided not only two lanes, but also pedestrian walkways – something really new.

The 1948 bridge was destroyed in 1964 by the flood caused by the failure of Hell Hole Dam upstream. The bridge you see today was built in 1965 in the same spot as the 1948 bridge. As recently as 1997, after a particularly heavy period of rain, the river water lapped the lower steel beams of this bridge.

Plants: The exposed areas beneath this stop are within the scour line of the river's floodwaters. Here are found many durable varieties of native plants including coffeeberry, redbud, buckwheat, styrax, toyon, philadelphus and locust.

Geology: The rivers in the confluence area make a series of right angle turns. These abrupt changes in direction are often the result of water following old fault zones that are made of crushed rock and thus easier to erode.

This stop completes the first loop section of this trail (Stops #1 to #4).

Walk across the Hwy 49 Bridge, turn right and continue back to the kiosk at the NW end of the Old Foresthill Bridge. Then cross the Old Foresthill Bridge (curved bridge) to the NE side. The next stop (#5) is on the left about 20 yards beyond the gate and the sign for the Lake Clementine Trail.

Marker Post #5
The Tallest Bridge in California

History: Up river above the North Fork is the Foresthill Bridge, finished in 1973. This bridge was built at a cost of 13.5 million dollars. The bridge would have spanned the lake created by the now stalled Auburn Dam. The lake would have inundated over 40 miles of canyons along the North and Middle Forks. Construction at the dam site was stopped in the late 1970's.

In 1973, the Foresthill Bridge was just nearing completion. At 730 feet high, it is the highest bridge in California and the 3rd highest in the United States.

The bridge has proven a favorite site for movies, TV shows and commercials as well as stunts - both legal and illegal. The movie *Breakdown* used the location for one scene and the action thriller *XXX* depicted a Corvette plunging off the bridge -- with the driver parachuting to safety.

Plants: The term "riparian" is used to describe vegetation along the river's edge. Along the right bank looking upstream you can see a riparian forest including: cottonwoods, willows, alders and valley oaks.

Geology: The flow of the North Fork American River below you is uncontrolled (as contrasted with the dam-controlled releases on the Middle Fork). This allows for very warm, low-flow conditions (and good swimming) in late summer. Look at the sizes of cobbles and boulders in the riverbed. Cobbles are rocks that are baseball to football in size while boulders are bigger rocks. Think about how fast the river needs to flow in order to move these rocks – an indication of the power of historic floods.

Marker Post #6
Fire, Pines and Birds

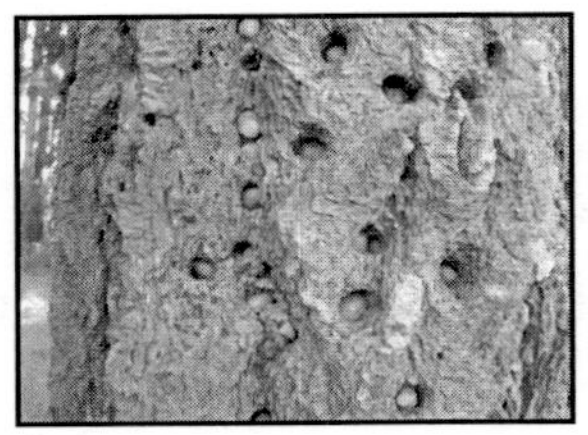

Acorn granary.

Plants: Because of frequent fires in this area you'll notice that there are fewer large trees and more open areas. Although most of these fires have been human-caused, the frequency approximates the natural occurrence of fire in the area and the vegetation patterns that result. The large ponderosa pines above the road are an important habitat for wildlife. The local acorn woodpeckers use several of the trees here to create granaries (acorn caches in the bark of the trees).

Marker Post #7
Steel Bowstring Truss, What?

History: Prior to the steel bowstring truss bridge erected in 1911, several low-level wooden bridges were built to offer a dry crossing to travelers. Looking across the river, you can see the concrete foundation for the steel bridge. Notice the heavily forested hillside in the photo (compared with the sparser growth today).

This bridge linking Auburn and Foresthill was reinforced in 1944 to carry heavier loads. Used extensively for logging traffic, it became inadequate for the increasingly larger vehicles. By 1952, the bridge had become the #1 deficiency of the county primary road system. So in 1954, construction started on the current bridge just downstream (known today as the Old Foresthill Bridge, to distinguish it from the tall Foresthill Bridge upstream). A two lane, state of the art bridge at the time, it was completed on schedule in 1955 and has withstood occasional severe high waters since then.

Plants: A riparian area and seasonal creek stimulates the larger trees and other vegetation in this area. Interior live oak, canyon live oak, black willow, white alder, and valley oak all grow within a 50-foot radius of this location.

The concrete abutments of the 1911-1955 steel bridge can still be seen today.

Compare the two types of blackberry bushes growing just a short distance further up the trail. The dominant one is the Himalayan blackberry, an introduced variety, which (in the mature form) has groups of five leaves, recurved spines and stems with ridges. The native blackberry has groups of three leaves, smaller thorns and smooth, gray-white stems. Both produce edible berries and habitat for birds and other animals.

Marker Post #8
How High Would the Water Be?

History: The Foresthill Bridge is the third highest in the U.S. and the highest in California. It is 2,248 feet in length and the bridge deck is 730 feet above the river. If the proposed dam was completed and the reservoir filled, the high-water level would hit just 22 feet below the top of the cement piers at over 590 feet above the river. The bridge opened on September 1,1973, with an elaborate ceremony involving an elephant and a donkey representing the two major political parties.

Plants: Notice the heavy tree cover on the opposite side of the canyon. How many different species can you see? (Refer to list included at Stop #3.)

Geology: Look for a white quartz vein in a boulder by the left side of the trail. This type of vein commonly occurs with gold deposits and is not present in the rocks at the confluence, indicating that this boulder was washed down river from its original location.

The high-water mark of the proposed Auburn Reservoir would have been about 590 feet above the river (see arrow), near the top of the concrete abutments of the high Foresthill Bridge.

Marker Post #9
Ready For A Swim?

Plants: Notice the difference in vegetation between the uphill and downhill sides of the roadway. In the undisturbed uphill area you can see dense tree cover and a thick under story of vegetation. The disturbed downhill area is much hotter and drier and supports a more limited variety and quantity of vegetation.

About one mile from the start of this walk you will come to a long, deep section of still water that extends for nearly ½ mile upstream, known locally as Clarks Hole. This pool was formed from an underwater dam of stacked cobbles and boulders, the result of early placer mining activity. The easily accessible rock-lined swimming hole has been popular with locals for over 100 years.

Clarks Hole and the 730 foot high Foresthill Bridge.

The Foresthill covered bridge was in use from 1875 to 1911, when it was replaced by a steel truss bridge further down stream.

Marker Post #10
Covered Bridge

History: Across the river you can clearly see remnants of the stone foundations of a wooden covered toll bridge. It was built in 1875 and used until 1911 for traffic between Auburn and Foresthill. Users were charged a toll of 6 cents for each cow, 50 cents for a horseman and $1 for a wagon and two horses.

A little further upstream, if you look carefully, you can see on the opposite bank the remains of the foundations of the three bridges that existed here between 1852 and 1875.

Geology: The straight lines and big flat surfaces in the rocks around Clarks Pool are natural fractures that have broken the rock – indications of faulting in the area millions of years ago.

Note: By continuing upstream another 0.5 miles on the Lake Clementine Trail (see separate trail guide) along a largely shaded roadway you will reach the paved Lake Clementine Rd. From there it is only a short ¼ mile downhill stroll to views of the North Fork Dam and to the Lake Clementine boat launching area.

Detailed Trail Index

Trail Name	From/To
American Canyon Trail	American Canyon to Western States Trail, (WST)
American Canyon - Dead Truck - WST	American Canyon to American Canyon
Applegate to Lake Clementine Trail	Boole Rd to Lake Clementine
Clark's Hole Trail	Stagecoach to NF American River
Codfish Falls Trail	Ponderosa Way to Codfish Falls
Confluence Trail	Mammoth Bar Rd to Old Foresthill Rd
Culvert Trail	Fuel Break Trail to Old Foresthill Rd
Foresthill Divide Loop	Circumnavigates Foresthill Rd
Fuel Break Trail	Clementine Rd to Foresthill Rd
Indian Creek Trail	Yankee Jims Rd to Indian Creek
Lake Clementine Access Trail	Clementine Rd to Lake Clementine
Lake Clementine Trail	Old Foresthill Rd to Clementine Rd
Lakeview Connector Trail	Clementine Rd to Foresthill Divide Loop/Rd
Olmstead Loop Trail	Knickerbocker Flats at Cool
Park HQ to Confluence Loop	Park HQ to Confluence to Park HQ
PG&E Trail	Hwy 49 to Quarry Road Trail
Pointed Rocks Trail	WST to Olmstead Loop Trail
Quarry Road Trail	Hwy 49 to Poverty Bar
Quarry & WST Loop	Hwy 49 to Hwy 49
Robie Point Firebreak Trail	Pacific St to Hwy 49
Salt Creek Loop Trail	Knickerbocker Flats in Cool to American River
Stagecoach Trail	Russell Rd to Old Foresthill Rd
Stevens Trail	Colfax to NF American River
Tamaroo Bar Trail	Overlook Park in Auburn to American River
Tinker's Cut Off Trail	Stagecoach to Hwy 49
Upper Stagecoach Trail	Stagecoach to Flood Rd Trail
WST - Hwy. 49 to Robie FB	Hwy 49 to Robbie Point Fuel Break Loop
Windy Point Trail	Iowa Hill Rd to NF American River
Confluence Self-Guided Interpr. Trail	Circumnavigates confluence area

Work on the Mt. Quarries Railroad bed in 1911. It is now part of the Auburn SRA trail system.

Mileage	Rating	Bike	Equestrian	Hike	ASRA Topo Map #	Fee	Page
2.4 M - one way	M to D		X	X	1		4
5.0 m - loop	E to M		X	X	1-10-45		6
2.5 M - one way	E to M	X	X	X			8
0.6 M - one way	E to M			X	5	$	10
1.7 M – one way	E			X	7	$	12
1.8 M - one way	M	X		X	8	$	14
1.0 M - one way	M	X		X	9		16
8.2 M - loop	E to M	X	X	X	12	$	18
1.5 M - one way	E to M	X		X	14		20
2.0 M - one way	E			X	15	$	22
1.5 M - one way	E to M	X		X	17		24
1.9 M - one way	E	X	X	X	18	$	26
3.6 M - one way	E to M	X	X	X			28
8.6 M - loop	E to M	X	X	X	26	$	30
3.5 M - loop	E to M			X	28-42-38-36-22		32
3.6 M - one way	E to M	X		X	29	$	34
1.6 M - one way	M to D		X	X	30		36
5.6 M - one way	E	X	X	X	31	$	40
6.3 M - loop	M to D		X	X	31-44-43-34	$	42
3.6 M - one way	E to M	X	X	X	33		44
7.5 M - loop	M to D	X	X	X		$	46
2.0 M - one way	E to M	X	X	X	36	$	48
3.2 M - one way	E to M	X		X	37		50
2.3 M - one way	E to M		X	X			52
0.3 M - one way	M to D			X	38		54
1.0 M - one way	E to M	X		X	39		56
2.0 M - one way	E to M		X	X			58
1.5 M - one way	M to D			X	49		60
2.6 M - loop	E			X		$	62

$ indicates current fee area - subject to change

River use just downstream of the Ponderosa Way Bridge near the Codfish Falls trailhead.

The Auburn State Recreation Area
One of California's State Parks

In the heart of the gold country, the Auburn State Recreation Area (Auburn SRA) is made up of over 38,000 acres, along 40 miles of the North and Middle Forks American River. Once teeming with thousands of gold miners, the area is now a natural area offering a wide variety of recreation opportunities to nearly a million visitors a year.

Major recreational uses include hiking, swimming, boating, fishing, camping, mountain biking, gold panning and off-highway motorcycle riding. Whitewater recreation is also very popular on both forks of the river, with class II, III, and IV runs. Nearly 40 private outfitters are licensed to offer whitewater trips in Auburn SRA.

More detailed information on all aspects of Auburn SRA is available from: 501 El Dorado St., Auburn, CA 95603 or www.parks.ca.gov.

The confluence of the North and Middle Forks of the American River is one of the most popular use areas for swimming, sunbathing and access to many of the trails in the Recreation Area.

Accessibility Information

To use the California Relay Service with TTY, call (888) 877-5378 or without TTY, call (888) 877-5379. Prior to arrival, visitors with disabilities who need special assistance should contact, (530) 885-4527. This publication is available in alternate formats by calling (530) 885-4527.

Thousands of gold miners took out millions of dollars of gold from the American River, starting in 1849. After a few years, individual efforts gave way to more organized gold mining operations, like this dredge that operated on the Middle Fork.